THIS WILL CHANGE
EVERYTHING

ALSO BY JOHN BROCKMAN

AS AUTHOR
By The Late John Brockman
37
Afterwords
The Third Culture: Beyond the Scientific Revolution
Digerati

AS EDITOR
About Bateson
Speculations
Doing Science
Ways of Knowing
Creativity
The Greatest Inventions of the Past 2,000 Years
The Next Fifty Years
The New Humanists
Curious Minds
My Einstein
Intelligent Thought

AS COEDITOR
How Things Are

THE *EDGE* SERIES
The Greatest Inventions of the Past 2,000 Years
What We Believe But Cannot Prove
What Is Your Dangerous Idea?
What Are You Optimistic About?
What Have You Changed Your Mind About?

THIS WILL CHANGE
EVERYTHING

IDEAS THAT WILL SHAPE THE FUTURE

EDITED BY JOHN BROCKMAN

HARPER  PERENNIAL

NEW YORK • LONDON • TORONTO • SYDNEY • NEW DELHI • AUCKLAND

FIRST EDITION

Designed by Rosa Chae

Library of Congress Cataloging-in-Publication data is available upon request.

ISBN 978-0-06-189967-6

10 11 12 13 14   OV/RRD   10 9 8 7 6 5 4 3 2 1

ACKNOWLEDGMENTS

———◦———

I wish to thank Peter Hubbard of HarperCollins for his encouragement.

I am also indebted to my agent, Max Brockman, who recognized the potential for this book, and to Sara Lippincott for her thoughtful and meticulous editing.

CONTENTS

PREFACE

---◄○►---

THE *EDGE* QUESTION

In 1991, I suggested the idea of a third culture which "consists of those scientists and other thinkers in the empirical world who, through their work and expository writing, are taking the place of the traditional intellectual in rendering visible the deeper meanings of our lives, redefining who and what we are." By 1997, the growth of the Internet had allowed implementation of a home for the third culture on the Web, on a site named *Edge* (www.edge.org).

Edge is a celebration of the ideas of the third culture, an exhibition of this new community of intellectuals in action. They present their work and their ideas, and they comment on the work and ideas of third-culture thinkers. They do so with the understanding that they are to be challenged. What emerges is rigorous discussion concerning crucial issues of the digital age, in a highly charged atmosphere where "thinking smart" prevails over the anesthesiology of wisdom.

The ideas presented on *Edge* are speculative; they represent the frontiers in such areas as evolutionary biology, genetics, computer science, neurophysiology, psychology, and physics. Some of the fundamental questions posed are: Where did the universe come from? Where did life come from? Where did the mind come from? Emerging out of the third culture are a new natural philosophy, new ways of understanding physical systems, new ways of thinking that call into question many of our basic assumptions of who we are, of what it means to be human.

An annual feature of *Edge* is The World Question Center, which was introduced in 1971 as a conceptual art project by my friend and collaborator the late artist James Lee Byars. His plan was to gather the hundred most brilliant minds in the world together in a room, lock them in, and "have them ask each other the questions they were asking themselves." The result was to be a synthesis of all thought.

Between idea and execution, however, are many pitfalls. Byars identified his hundred most brilliant minds, called each of them, and asked them what questions they were asking themselves. The result: Seventy people hung up on him.

But by 1997, the Internet and e-mail had allowed for a serious implementation of Byars's grand design, and this resulted in launching *Edge*. For each of the anniversary editions of *Edge*, I have used the interrogative myself and asked contributors for their responses to a question that comes to me, or to one of my correspondents, in the middle of the night.

New tools equal new perceptions. Through science we create technology and in using our new tools we re-create ourselves. But until very recently in our history, no democratic populace, no legislative body, ever indicated by choice, by vote, how this process should play out.

Nobody ever voted for printing. Nobody ever voted for electricity. Nobody ever voted for radio, the telephone, the automobile, the airplane, television. Nobody ever voted for penicillin, antibiotics, the Pill. Nobody ever voted for space travel, massively parallel computing, nuclear power, the personal computer, the Internet, e-mail, cell phones, the Web, Google, cloning, sequencing the entire human genome. We are moving toward the redefinition of life, to the edge of creating life itself. While science may or may not be the only news, it is the news that stays news. And our politicians, our governments? Always years behind, the best they can do is play catch-up.

Nobel laureate James Watson, codiscoverer of the DNA double helix, and genomics pioneer J. Craig Venter recently received Double Helix Awards from Cold Spring Harbor Laboratory as the founding fathers of human-genome sequencing. They are the first two human beings to have their complete genetic information decoded. Watson noted in his acceptance speech that he doesn't want government involved in decisions concerning how people choose to handle information about their personal genomes. Venter is on the brink of creating the first artificial lifeform on Earth. He has already announced transplanting the information from one genome into another; in other words, your dog becomes your cat. He has privately alluded to important scientific progress in his lab, the result of which, if and when realized, will change everything.

Here is the 2009 *Edge* question:

WHAT WILL CHANGE EVERYTHING?

"What game-changing scientific ideas and developments do you expect to live to see?"

JOHN BROCKMAN
PUBLISHER AND EDITOR, *EDGE*

INTRODUCTION

---◦---

DANIEL C. DENNETT

DANIEL C. DENNETT is a philosopher, university professor, codirector of the Center for Cognitive Studies at Tufts University, and the author of *Breaking the Spell: Religion as a Natural Phenomenon.*

What will change everything? The *Edge* question itself, and many of the answers given here, point to a common theme: Reflective, scientific investigation of everything is going to change everything. When we look closely at looking closely, when we increase our investment in techniques for increasing our investment in techniques . . . for increasing our investment in techniques, we create nonlinearities that—like Douglas Hofstadter's strange loops—amplify uncertainties, allowing phenomena that have heretofore been orderly and relatively predictable to escape our control. We figure out how to game the system, and this initiates an arms race to control or prevent the gaming of the system, which leads to new levels of gamesmanship, and so on.

The snowball has started to roll and there is probably no stopping it. Will the result be a utopia or a dystopia? Which of the novelties are self-limiting and which will extinguish institutions long thought to be permanent? There is precious little inertia, I think, in cultural phenomena once they are placed in these arms races of cultural evolution. Extinction can happen overnight in some cases. The almost frictionless markets made possible by the Internet are already swiftly revolutionizing commerce.

Will universities and newspapers become obsolete? Will hospitals and churches go the way of corner grocery stores and livery stables? Will reading music soon become as arcane a talent as reading hieroglyphics? Will reading and writing themselves soon be obsolete? What will we use our minds for? Some see a revolution in our concept of intelligence, either because of "neurocosmetics" (Marcel Kinsbourne) or quantum computing (W. H. Hoffman), or "just-in-time storytelling" (Roger Schank). Nick Humphrey reminds us that when we get back to basics—procreating, eating, just staying alive—not that much has changed since Roman times, but I think that these are not really fixed points after all.

Our species' stroll through Design Space is picking up speed. Recreational sex, recreational eating, and recreational perception (hallucinogens, alcohol) have been popular since Roman times, but we are now on the verge of recreational self-transformations that will dwarf the modifications the Romans indulged in. When you no longer need to eat to stay alive, or procreate to have offspring, or locomote to have an adventure-packed life, when the residual instincts for these activities might be simply turned off by genetic tweaking, there may be no constants of human nature left at all. Except, maybe, our incessant curiosity.

EVOLUTION CHANGES EVERYTHING

—◦—

SCOTT SAMPSON

SCOTT SAMPSON is adjunct associate professor of geology and geophysics at the University of Utah and the host of *Dinosaur Planet*.

Evolution is the scientific idea that will change everything within the next several decades.

I realize that this statement may seem improbable. If evolution is defined generally simply as change over time, the above statement borders on meaningless. If it is regarded in the narrower, Darwinian sense as descent with modification, any claim for evolution's starring role also appears questionable—particularly given that 2009 is the hundred and fiftieth anniversary of the publication of *On the Origin of Species*. Surely what Daniel Dennett has called Darwin's "dangerous idea," however conceived, has made its mark by now. Nevertheless I base my claim on evolution's probable effects in two great spheres: human consciousness, and science and technology.

Today the commonly accepted conception of evolution is extremely narrow, confined largely to the realm of biology and a longstanding emphasis on mutation and natural selection. In recent decades, this limited perspective has become further entrenched by the dominance of molecular biology and its "promise" of human-engineered cells and life-forms. Emphasis has been placed almost entirely on generating diversity, a process referred to as "complexification," reflecting the reductionist worldview that has driven science for four centuries.

Yet science has also begun to explore another key element of evolution: unification, which transcends the biological to encompass the evolution of physical matter. The numerous and dramatic increases in complexity, it turns out, have been achieved largely through a process of integration, with smaller wholes becoming parts of larger wholes. Again and again, we see the progressive development of multipart individuals from simpler forms. Thus, for example, atoms become integrated into molecules, molecules into cells, and cells into organisms. At each higher, emergent stage, older forms are enveloped and incorporated into newer forms, with the end result being a nested, multilevel hierarchy.

At first glance, the process of unification appears to contravene the second law of thermodynamics, by increasing order over entropy. Again and again during the past fourteen billion years, concentrations of energy have emerged and self-organized as islands of order amid a sea of chaos, taking the guise of stars, galaxies, bacteria, gray whales, and (on at least one planet) a biosphere. Although the process of emergence remains somewhat of a mystery, we can now state with confidence that the epic of evolution has been guided by counterbalancing trends of complexification and unification. This journey has not been an inevitable, deterministic march but a quixotic, creative unfolding in which the future could not be predicted.

How will a more comprehensive understanding of evolution affect science and technology? Already a nascent but fast-growing industry called biomimicry taps into nature's wisdom, imitating sustainable, high-performance designs and processes built up during 4 billion years of evolutionary R&D. Water-repellent lotus plants inspire nontoxic fabrics. Termite mounds inspire remarkable buildings that make use of passive cooling. Spider silk may provide inspiration for a new, strong, flexible yet rigid material with innumerable possible uses. Ultimately, plant photosynthesis may reveal secrets to tapping an unlimited energy supply with minimal waste products.

The current bout of biomimicry is just the beginning. I am increasingly convinced that ongoing research into such phenomena as complex adaptive systems will result in a new synthesis of evolution and energetics—let's call it the Unified Theory of Evolution—that will trigger a cascade of novel research and designs. Science will relinquish its unifocal downward gaze on reductionist nuts and bolts and turn upward to explore the "pattern that connects." An understanding of complex adaptive systems will yield transformative technologies we can only begin to imagine. Think about the potential for new generations of "smart" technologies, with the ability to adapt—indeed, to evolve and transform—in response to changing conditions.

And what of human consciousness? Reductionism has yielded stunning advances in science and technology; however, its dominant metaphor, life-as-machine, has left us with a gaping chasm between the human and nonhuman worlds. With "nature" (the nonhuman world) reduced merely to resources, humanity's ever-expanding activities have become too much for the biosphere to absorb. We have placed ourselves and the biosphere on the precipice of a devastating ecological crisis, without the consciousness for meaningful progress toward sustainability.

At present, Western culture lacks a generally accepted cosmology—a story that gives life meaning. One of the greatest contributions of the scientific enterprise is the epic of evolution. For the first time, thanks to the combined efforts of astronomers, biologists, and anthropologists (among many others), we have a realistic, time-developmental understanding of the fourteen-billion-year history of us. Darwin's tree of life has roots that extend back to the Big Bang, and fresh green shoots reach into an uncertain future. Far from leading to a view that the universe is meaningless, this saga provides the foundation for seeing ourselves as fully embedded in the fabric of nature. To date, this story has had minimal exposure and certainly has not been included (as it should be) in the core of our educational curricula.

Why am I confident that these transformations will occur in the near future? In large part because necessity is the mother of invention. We are the first generation of humans to face the prospect that humanity may have a severely truncated future. In addition to new technologies, we need a new consciousness, a new worldview, and new metaphors that establish a more harmonious relationship between the human and the nonhuman. Of course, the concept of "changing everything" makes no up-front value judgments, and I can envision evolution's net contribution as being either positive or negative, depending on whether the shift in human consciousness keeps pace with the radical expansion of new (and potentially even more exploitative) technologies. In sum, our future R&D needs to address human consciousness in at least equal measure to science and technology.

DNA: WRITING THE SOFTWARE OF LIFE

——◁◦▷——

J. CRAIG VENTER

J. CRAIG VENTER is a geneticist, founder and president of the J. Craig Venter Institute, and the author of *A Life Decoded: My Genome, My Life*.

In science, as with most areas, seemingly simple ideas can change and have changed everything. Just one hundred and fifty years ago, Charles Darwin's *On the Origin of Species* was published and immediately affected science and society by describing the process of evolution as natural selection. But it took until the 1940s to establish that the substance carrying the heritable information was DNA. In 1953, an Englishman and an American, Francis Crick and James Watson, proposed that DNA is formed as a spiraling ladder—or double helix—with purine-pyrimidine base pairs as the rungs. However, no one yet knew what the actual code of life was.

In the 1960s, some of the first secrets of our genetic code were revealed with the discovery that the chemical bases should be read in groups of three. These "nucleotide triplets" then coded for amino acids, which form protein.

In the late 1970s, the complete genetic code (five thousand nucleotides) of bacteriophage, a small virus that kills *E. coli*, was read out in sequence by a new technology developed by Fred Sanger from the University of Cambridge. This technology, named Sanger sequencing, would dominate genetics for the next twenty-five years.

In 1995 my team at the Institute for Genomic Research (now the J. Craig Venter Institute) read the complete genetic code of the chromosome containing all of the genetic information for a bacterium. The bacterial genome we decoded was over 1.8 million nucleotides long and coded for all the proteins associated with the life of the bacterium. Based on our new methods, there was an explosion of new data from decoded genomes of many living species, including humans.

Just as Darwin observed evolution in the changes he saw in various species of finches, land and sea iguanas, and tortoises, the genomics community is now studying the changes in the genetic code that are associated with human traits and disease and the differences among us by reading the genetic code of many humans and comparing them. The technology is changing rapidly; soon it will be commonplace for everyone to know his or her own genetic code. This will change the practice of medicine from treating disease after it happens to preventing disease before its onset. Understanding the mutations and variations in the genetic code clearly will help us to understand our own evolution.

Science is changing dramatically once again, as we use all our new tools to understand life and perhaps even to redesign it. The genetic code is the result of over 3.5 billion years of evolution and is common to all life on our planet. We have been reading the genetic code for a few decades and are gaining insight into how it programs for life. In a series of experiments to better understand the code, my colleagues and I have developed new ways to chemically synthesize DNA in the laboratory. First, we synthesized the genetic code of the same virus that Sanger decoded in 1977. When this large synthetic molecule was inserted into a bacterium, not only was the cellular machinery in the bacterium able to read the synthetic genetic code, but the cell was also able to produce the proteins that were coded for. The proteins self-assembled to produce the virus particle, which was then able to infect other bacteria. Over the past few years, we have been able to chemically make an entire bacterial chromosome, which, at more than 582,000 nucleotides, is the largest man-made chemical produced to date.

We have now shown that DNA is absolutely the information-coded material of life by completely transforming one species into another simply by changing the DNA in the cell. By inserting a new chromosome into a cell and eliminating the existing chromosome, all the characteristics of the original species were lost and replaced by what was coded for on the new chromosome. Very soon, we will be able to do the same experiment with the synthetic chromosome.

We can start with digitized genetic information and four bottles of chemicals and write new software of life to direct organisms to do processes that are desperately needed, like the creation of renewable biofuels and the recycling of carbon dioxide. As we learn from 3.5 billion years of evolution, we will convert billions of years into decades and change not only how we view life but life itself.

A CHANGE IN WHO WE ARE

◀◦▶

PZ MYERS

PAUL ZACHARY (PZ) MYERS is an associate professor of biology at the University of Minnesota and the author of the science blog *Pharyngula*.

The question "What will change everything?" is framed in the wrong tense. It should be "What is changing everything right now?"

We're in the midst of an ongoing revision of our understanding of what it means to be human. We are struggling to redefine humanity, and it's going to radically influence our future. The redefinition began in the nineteenth century with the work of Charles Darwin, who changed the game by revealing the truth of human history: We are not the progeny of gods, we are the children of worms; not the product of divine planning but of cruel chance and ages of brutal winnowing. That required a shift in the way we view ourselves—a shift that is still working its way through the culture; creationism is an instance of the reaction against the dethroning of *Homo sapiens*. Embracing the perspective of evolution, however, allows us to see the value of other species and appreciate our place in the system as a whole; it is a positive advance.

There are at least two more revolutions in the works. The first is in developmental biology. We're learning how to reprogram human tissues, enabling new possibilities in repair and regeneration. We're acquiring the tools that will make the human form more plastic, and it won't just stop with restoring damaged bodies but will someday allow us to resculpt ourselves, adding new features and properties to our biology, and—maybe some-day—free us completely from the boundaries of the fixed form of a bipedal primate. Even now, with our limited abilities, we have to rethink what it means to be human. Does a blastocyst fit the definition? How about a five-week-old embryo or a three-month-old fetus?

The second big revolution is coming from neuroscience. Mind is clearly a product of the brain, and the old notions of souls and spirits are looking increasingly ludicrous. Yet these are nearly universal ideas, all tangled up in people's rationalizations about an afterlife and ultimate reward or punishment and in their concept of self. If many object to the lack of exceptionalism in our history, if they're resistant to the idea that human identity emerges gradually during development, they're most definitely going to find the idea of soullessness and mind as a by-product of nervous activity horrifying.

This will be our coming challenge: to accommodate a new view of ourselves and our place in the universe that isn't encumbered by falsehoods and trivializing myths. That's going to be our biggest change: a change in who we are.

THE ROBOTIC MOMENT

◄○►

SHERRY TURKLE

SHERRY TURKLE is a psychologist at MIT and the author of *Evocative Objects: Things We Think With.*

I will see the development of robots that people will want to spend time with. Not just a little time—in which the robots serve as amusements—but enough time and with enough interactivity that the robots will be experienced as companions, each closer to a someone than a something. I think of this as the robotic moment.

Sociable technologies first came on the mass market with the 1997 Tamagotchi, a creature on a small video screen that did not offer to take care of you but asked you to take care of it. The Tamagotchi needed to be fed and amused. It needed its owners to clean up after its digital messes. Tamagotchis demonstrated that in digital sociability, nurturance is a "killer app." We nurture what we love, but we love what we nurture. In the early days of artificial intelligence, the emphasis had been on building artifacts that impressed with their knowledge and understanding. When AI goes sociable, the game changes. The "relational" artifacts that followed the Tamagotchis inspired feelings of connection because they pushed people's Darwinian buttons: They asked us to teach them, they made eye contact, they tracked our motions, they remembered our names. For people, these are the markers of sentience; they signal us, rightly or wrongly, that there is "somebody home."

Sociable technologies came onstage as toys, but in the future they will be presented as potential nannies, teachers, therapists, life coaches, and caretakers for the elderly. First, they will be put forward as "better than nothing." (It is better to have a robot as a diet coach than just to read a diet book. If your mother is in a nursing home, it is better to leave her interacting with a robot that knows her habits and interests than leave her staring at a television screen.) But over time, robots will be presented as "better than something": That is, they will be preferable to an available human being—or, in some cases, a living pet. They will be promoted as having powers—of memory, attention, and patience—that people lack.

Even now, when people learn that I work with robots, they tell me stories of human disappointment: They talk of cheating husbands, wives who fake orgasms, children who take drugs. They despair about human opacity: "We never know how another person really feels; people put on a good face. Robots would be safer." As much as a story of clever engineering, our evolving attachments to technology speak to feelings of unrequited love.

In the halls of a large psychology conference, a graduate student takes me aside to ask for more information on the state of research about relational machines. She confides that she would trade in her boyfriend for "a sophisticated Japanese robot," if the robot would produce what she terms "caring behavior." She tells me that she relies on "a feeling of civility in the house." She does not want to be alone. She says: "If the robot could provide the environment, I would be happy to help produce the illusion that there is somebody really with me." What she is looking for, she tells me, is a "no-risk relationship" that will stave off loneliness. A responsive robot, even if it is just exhibiting scripted behavior, seems better to her than a demanding boyfriend. I ask her if she is joking. She tells me she is not.

A reporter for *Scientific American* calls to interview me about a book on robot love by computer scientist David Levy. In *Love + Sex with Robots*, Levy argues that robots will teach us to be better friends and lovers, because we will be able to practice on them relationally and physically. Beyond this, they can substitute where people fail us. Levy proposes, among other things, the virtues of marriage to robots. He argues that robots are "other" but in many ways better. No cheating. No heartbreak.

I tell the reporter that I am not enthusiastic about Levy's suggestions. To me, the fact that we are discussing marriage to robots is a window onto human disappointments. The reporter asks if my opposition to people marrying robots doesn't put me in the same camp as those who for so long stood in the way of marriage for lesbians and gay men. I try to explain that just because I don't think people should marry machines doesn't mean that any mix of adult people with other adult people isn't fair territory. He accuses me of species chauvinism and restates his objection: Isn't this the kind of talk that homophobes once used—not considering gays and lesbians as "real people"? Machines are "real" enough to bring special pleasures to relationships, pleasures that need to be honored in their own right.

The argument in *Love + Sex* is exotic, but we are being prepared for the robotic moment every day. Consider Joanie, age seven, who has been given a robot dog. She can't have a real dog because of her allergies, but the robot's appeal goes further. It is not just better than nothing but better than something. Joanie's robot, known as an Aibo, is a dog that can be made to measure. Joanie says, "It would be nice to be able to keep Aibo at a puppy stage for people who like to have puppies."

It is a very big step from Joanie's admiring a forever-young Aibo to David Levy and his robot lover. But they share the fantasy that while we may begin with substituting a robot if a person is not available, we will move on to specifically choosing malleable artificial companions. If the robot is a pet, it might always stay a puppy because that's how you like it. If the robot is a lover, you might always be the center of its universe because that's how you like it.

But what will happen if we get what we say we want? If our pets always stay puppy-cute; if our lovers always say the sweetest things? If you know only cute and cuddly, you don't learn about maturation, growth, change, and responsibility. If you know only an accommodating partner, you end up knowing neither the partner nor yourself.

The robotic moment will bring us to the question we must ask of every technology: Does it serve our human purposes?—a question that causes us to reconsider what those purposes are. When we connect with the robots of the future, we will tell and they will remember. But have they listened? Have we been "heard" in a way that matters? Will we no longer care?

THE BRAIN-MACHINE INTERFACE

―――――◇―――――

JAMES GEARY

JAMES GEARY is former Europe editor of *Time* and the author of *Geary's Guide to the World's Great Aphorists.*

J. Craig Venter may be "on the brink of creating the first artificial life-form," but one game-changing scientific idea I expect to live to see is the moment when a robotic device achieves the status of a living thing. What convinces me of this is not some amazing technological breakthrough but videos of the annual RoboCup soccer tournament organized by Georgia Tech in Atlanta. The robotics researchers behind RoboCup are determined to build a squad of robots capable of winning against the world champion human soccer team. For now, they're just competing against other robots.

For a human being to raise a foot and kick a soccer ball is an amazingly complex event, involving millions of different neural computations coordinated across several different brain regions. For a robot to do it—and to do it as gracefully as members of the RoboCup Humanoid League—is a major technical accomplishment. The cuddlier, though far less accomplished, quadrupeds in the Four Legged League are also a wonder to behold. Plus, the robots are not programmed to do this stuff—they learn to do it, just like you and me.

These robots are marvels of technological ingenuity. They are also "living" proof of how easily—eagerly, even—we can anthropomorphize robots, and why I expect there won't be much of a fuss when these little metallic critters start infiltrating our homes, offices, and daily lives.

I also expect to see the day when robots like these have biological components (that is, some wetware to go along with their hardware) and when human beings have internal technological components (that is, some hardware to go along with our wetware). Researchers at the University of Pittsburgh have trained two monkeys to grab marshmallows using a robotic arm controlled by their own thoughts. During voluntary physical movements (such as reaching for food), nerve cells in the brain start firing well before any movement actually takes place. It's as if the brain warmed up for an impending action by directing specific clusters of neurons to fire, just as a driver warms up a car by pumping the gas pedal. The University of Pittsburgh team implanted electrodes in this area of the monkeys' brains and connected them to a computer operating the robotic limb. When the monkeys thought about reaching for a marshmallow, the mechanical arm obeyed that command. In effect, the monkeys had three arms for the duration of the experiments.

In humans, this type of brain-machine interface (BMI) could open up new fields of entertainment and exploration—and also allow paralyzed individuals to control prosthetic body parts. "The body's going to be very different a hundred years from now," Miguel Nicolelis, a Duke University neuroscientist and one of the pioneers of BMI, has said. "In a century's time, you could be lying on a beach on the east coast of Brazil, controlling a robotic device roving on the surface of Mars, and be subjected to both experiences simultaneously, as if you were in both places at once. The feedback from that robot billions of miles away will be perceived by the brain as if it was you up there."

In robots, a BMI could become a kind of mind. If manufacturers create such robots with big puppy-dog eyes—or even wearing the face of a loved one or a favorite film star—I think we'd grow to like them pretty quickly. When they have enough senses and "intelligence," then I'm convinced that these machines will qualify as living things. Not human beings, by any means, but kind of high-tech pets. And turning one off will be the moral equivalent of shooting your dog.

BREAKING THE SPECIES BARRIER

———◦———

RICHARD DAWKINS

RICHARD DAWKINS is emeritus professor of the Public Understanding of Science, University of Oxford, and the author of *The God Delusion* and *The Greatest Show on Earth*.

Our ethics and our politics assume, largely without question or serious discussion, that the division between human and "animal" is absolute. "Pro-life," to take just one example, is a potent political badge, associated with a gamut of ethical issues such as opposition to abortion and euthanasia. What it really means is pro-*human*-life. Abortion clinic bombers are not known for their veganism, nor do Roman Catholics show any particular reluctance to have their suffering pets "put to sleep." In the minds of many confused people, a single-celled human zygote, which has no nerves and cannot suffer, is infinitely sacred, simply because it is "human." No other cells enjoy this exalted status.

But such "essentialism" is deeply unevolutionary. If there were a heaven in which all the animals who ever lived could frolic, we would find an interbreeding continuum between every species and every other. For example, I could interbreed with a female who could interbreed with a male who could . . . fill in a few gaps, probably not very many in this case . . . who could interbreed with a chimpanzee. We could construct longer but still unbroken chains of interbreeding individuals to connect a human with a warthog, a kangaroo, a catfish. This is not a matter of speculative conjecture; it necessarily follows from the fact of evolution.

Theoretically, we understand this. But what would change everything is a practical demonstration, such as one of the following:

1. The discovery of relict populations of extinct hominins, such as *Homo erectus* or *Australopithecus*. Yeti enthusiasts notwithstanding, I don't think this is going to happen. The world is now too well explored for us to have overlooked a large, savannah-dwelling primate. Even *Homo floresiensis* has been extinct for seventeen thousand years. But if it did happen, it would change everything.

2. A successful hybridization between a human and a chimpanzee. Even if the hybrid were infertile, like a mule, the shock waves that would be sent through society would be salutary. This is why a distinguished biologist described this possibility as the most immoral scientific experiment he could imagine: It would change everything! It cannot be ruled out as impossible, but it would be surprising.

3. An experimental chimera in an embryology lab, consisting of approximately equal numbers of human and chimpanzee cells. Chimeras of human and mouse cells are now constructed in the laboratory as a matter of course, but they don't survive to term. Incidentally, another example of our species-ist ethics is the fuss now made about mouse embryos containing some proportion of human cells: "How human must a chimera be before more stringent research rules should kick in?" So far, the question is merely theological, since the chimeras don't come anywhere near being born, and there is nothing resembling a human brain. But, to venture

down the slippery slope so beloved of ethicists, what if we were to fashion a chimera of 50 percent human and 50 percent chimpanzee cells and grow it to adulthood? That would change everything. Maybe it will.

4. The human genome and the chimpanzee genome are now known in full. Intermediate genomes of varying proportions can be interpolated on paper. Moving from paper to flesh and blood would require embryological technologies that will probably come onstream during the lifetime of some of my readers. I think it will be done, and an approximate reconstruction of the common ancestor of ourselves and chimpanzees will be brought to life. The intermediate genome between this reconstituted "ancestor" and modern humans would, if implanted in an embryo, grow into something like a reborn *Australopithecus*: Lucy the Second. And that would (dare I say "will"?) change everything.

I have laid out four possibilities that would, if realized, change everything. I have not said that I hope any of them will be realized. That would require further thought. But I will admit to a *frisson* of enjoyment whenever we are forced to question the hitherto unquestioned.

SLIPPERY EXPECTATIONS

———◆———

COREY S. POWELL

COREY S. POWELL is editor in chief of *Discover* magazine.

The tricky, slippery word in the *Edge* question is "expect." There is nothing I expect to see with 100 percent certainty, and there are some remarkable things I expect to see with perhaps 10 or 5 percent certainty (but I sure would be excited if that 5 percent paid off). With that bit of preamble, I'd like to lay out my game-changing predictions, ranked by order of expectation, starting with the near-sure things and ending with the thrilling hey-you-never-knows.

The real end of oil. Technology will make liquid fuels obsolete—not just petroleum, but also alternatives like biodiesel, ethanol, and so on. Fossil fuel supplies are too volatile and limited, the fuels themselves are far too environmentally costly, and biofuels will never be more than niche players. More broadly, moving fuels around in liquid form is just too cumbersome. In the future, energy for personal transit might be delivered by wire or by beam. It might not be delivered at all: The *Back to the Future* "Mr. Fusion" device is not all that far-fetched (see below). But whatever comes next, in another generation or so pumping fuel into a car will seem as quaint as getting out and cranking the engine to get it started. *Odds: 95 percent.*

Dark matter found. The hunt for the Higgs boson is a yawn, from my perspective. Finding it will only confirm a theory (the

Standard Model of particle physics) that most physicists are fairly sure about already. Identifying dark-matter particles—either at the Large Hadron Collider (LHC) or at one of the direct detectors, like XENON100—would be much more significant. It would tell us what the other five-sixths of all matter in the universe consists of, it would instantly rule out a lot of kooky cosmological theories, and it would allow us to construct a complete history of the universe. *Odds: 90 percent.*

Genetically engineered kids. I'm talking not just about screening out major cancer genes or selecting blue eyes, I'm talking about designing kids who can breathe underwater or who have radically enhanced mental capabilities. Such offspring will rewrite the rules of evolution and redefine what it means to be human. They may very well qualify as totally new species. From a scientific point of view, I think this capability is extremely likely, but legal and ethical considerations may prevent it from happening. With all that, I put the odds at: *80 percent.*

Life detected on an exoplanet. Astronomers have already measured the size, density, temperature, and atmospheric composition of several alien worlds as they transit in front of their parent stars. The upcoming James Webb Space Telescope may be able to do the same for Earth-sized planets. We haven't found these planets yet, but we're getting close; it's a shoo-in that the recently launched Kepler Mission or one of the ground-based planet searches will find them soon. The real question is whether the chemical evidence of alien life will be conclusive enough to convince most scientists. (As for life on Mars, I'd say the odds are similar that we'll find evidence of fossil life there, but the likelihood of cross-contamination between Mars and Earth makes Martian life inherently less interesting.) *Odds: 75 percent.*

Synthetic telepathy. Rudimentary brain prostheses and brain-machine interfaces already exist. Allowing one person to control another person's body would be a fairly simple extension of that technology. Enabling one person to transmit thoughts directly to another person's brain is a much trickier proposition, but not terribly far fetched, and it would break down one of the most profound isolations associated with the human condition. Broadcasting the overall state or "mood" of a brain would probably come first. Transmitting specific, conscious thoughts would require elaborate physical implants to make sure the signals go to exactly the right place—but such implants could soon become common anyway, as people merge their brains with computer data networks. *Odds: 70 percent.*

Life span past two hundred (or a thousand). I have little doubt that progress in fighting disease and patching up our genetic weaknesses will make it possible for people to routinely reach the full human life span of about a hundred and twenty. Going far beyond that will require halting or reversing the core aging process, which involves not just genetic triggers but also oxidation and simple wear and tear. Engineering someone to have gills is probably a much easier proposition. Still, if we can hit two hundred, I see no reason why the same techniques couldn't allow people to live to a thousand or more. *Odds: 60 percent.*

Conscious machines. Intelligent machines are inevitable—by some measures they are already here. Synthetic consciousness would be a much greater breakthrough, in some ways a more profound one than finding life on other planets. One problem: We don't understand how consciousness works, so re-creating it will require learning a lot more about what it means to be both smart and self-aware. Another problem: We don't understand what consciousness *is*, so it's not clear what "smart" and "self-aware" mean, exactly. Neurobiologist Gerald Edelman's brain-based devices are a promising solution. Rather than trying to deconstruct the brain as a computer, they construct neural processing from the bottom up, mimicking the workings of actual neurons. *Odds: 50–50.*

Geoengineering. We may be able to deal with global warming through a combination of new energy sources, radical efficiency, carbon sequestration, and many local and regional adaptations to a warmer climate. All of these will be technologically challenging but not truly "game changing." It is possible, though, that the result of our environmental follies will be so severe, and the progress of curative scientific research so dramatic, that some of the pie-in-the-sky geoengineering schemes now being bandied about will actually come to pass. Giant space mirrors and sunshades strike me as the most appealing options, both because they would support an aggressive space program and because they are adjustable and correctable. (Schemes that aim to fight carbon pollution with sulfur pollution seem like a frightening mix of hubris and folly.) Geoengineering techniques are also a good first step toward being able to terraform other planets. *Odds: 25 percent.*

Desktop fusion. The ITER (International Thermonuclear Experimental Reactor) project will prove that it is possible to spend billions of dollars to construct an enormous device that produces controlled hydrogen fusion at a net loss of energy. A few left-field fusion researchers—most notably the ones associated with the Tri Alpha company in Southern California—are exploring a much wilder, albeit quite unproved, approach that would lead to the construction of cheap, compact reactors. These devices could in theory take advantage of more exotic neutron-free fusion reactions that would allow almost direct conversion of fusion energy to electricity. The old dream of a limitless power plant that could fit under the hood of your car or in a closet in your house might finally come true. Since energy is the limiting factor for most economic development, the world economy (and the potential for research and exploration) would be utterly transformed. *Odds: 20 percent.*

Communication with other universes. Studies of gravitational-wave patterns etched into the cosmic microwave background could soon provide hints of the existence of universes outside our own. Particle collisions at the LHC could soon provide hints of the existence of higher dimensions. But what would really shake the world would be direct measurements of other universes. How exactly that would work is not at all clear, since any object or signal that crossed over directly from another universe could have devastating consequences; indirect evidence, meanwhile, might not be terribly convincing (for example, looking for the gravitational pull from shadow matter on a nearby brane). I hold out hope all the same. *Odds: 10 percent.*

Antigravity devices. Current physics theory doesn't allow such things, but from time to time fringe experiments (mostly involving spinning superconducting disks) allegedly uncover evidence for an antigravity phenomenon. Even NASA has invested dribbles of money in this field, hoping that something exciting and unexpected will pop up. If antigravity really exists, it would require revising Einstein's general theory of relativity. (It would also vindicate all those science-fiction TV shows in which everyone clomps around heavily in outer space.) Given how little we know about how gravity works, neither antigravity nor artificially generated gravity seem impossible . . . just highly improbable. *Odds: 5 percent.*

ESP verified! Probably the closest thing I've seen to a scientific theory of ESP is Rupert Sheldrake's concept of "morphic fields." Right now, there's nary a shred of evidence to support the idea—unless you count anecdotes of dogs who know when their owners are about to return home and people who can "feel" when someone is looking at them—but Sheldrake is totally correct that such off-the-wall ideas merit serious scientific investigation. After all, scientists investigate counterintuitive physics concepts all the time; why not conduct equally serious investigations of the intuitive feelings that people have all the time? Everything I know about science and human subjectivity says that there's nothing to find here. And yet when I think of a discovery that would change everything, this is one of the first that springs to mind. *Odds: 0.1 percent.*

THE FULL FLOURISHING OF SOLAR TECHNOLOGY

IAN McEWAN

IAN McEWAN is a novelist and the author of *On Chesil Beach*.

Philip Larkin began a poem with the hypothesis, "If I were called in / To construct a religion / I should make use of water." Instead of water, I would propose the sun, and the religion I have in mind is a rational affair, with enormous aesthetic possibilities and of great utility.

By nearly all insider and expert accounts, we are or will be at peak oil somewhere between now and the next five years. Even if we did not have profound concerns about climate change, we would need to be looking for different ways to power our civilization. How fortunate we are to have a safe nuclear facility a mere ninety-three million miles away, and fortunate too that the dispensation of physical laws is such that when a photon strikes a semiconductor, an electron is released. I hope I live to see the full flourishing of solar technology—photovoltaics or concentrated solar power to superheat steam, or a combination of the two in concentrated photovoltaics. The technologies are unrolling at an exhilarating pace, with input from nanotechnology and artificial photosynthesis. Electric mobility and electricity storage are also part of this new quest. My hope is that architects will be drawn to designing gorgeous arrays and solar towers in the desert—as expressive of our aspirations as medieval cathedrals once were. We will need new distribution systems too, smart grids—perfect Rooseveltian projects for our hard-pressed times. Could it be possible that in two or three decades we will look back and wonder why we ever thought we had a problem when we are bathed in such a sweet rain of photons?

PERSONAL GENOMICS—OR MAYBE NOT

---◦---

STEVEN PINKER

STEVEN PINKER is Johnstone Family Professor in the Department of Psychology, Harvard University, and the author of *The Stuff of Thought: Language as a Window into Human Nature*.

I have little faith in anyone's ability to predict what will change everything. A look at the futurology of the past turns up many chastening examples of confident predictions of technological revolutions that never happened, such as domed cities, nuclear-powered cars, and meat grown in dishes. By the year 2001, according to the eponymous movie, we were supposed to have suspended animation, missions to Jupiter, and humanlike mainframe computers (though not laptop computers or word processing—the characters used typewriters). And remember interactive television, the Internet refrigerator, and the paperless office?

Technology may change everything, but it's impossible to predict how. Take another way in which *2001: A Space Odyssey* missed the boat. The American women in the film were "girl assistants": secretaries, receptionists, and flight attendants. As late as 1968, when Kubrick's movie came out, few people foresaw the second feminist revolution that would change everything in the 1970s. It's not that the revolution didn't have roots in technological change. Not only did oral contraceptives make it possible for women to time their childbearing, but a slew of earlier technologies (sanitation, mass production, modern medicine, electricity) had reduced the domestic workload, extended the life span, and shifted the basis of the economy from brawn to brains, collectively emancipating women from round-the-clock child rearing.

The effects of technology depend not just on what the gadgets do, but on billions of people's judgments of their costs and benefits. (Do you really want to have to call a help line to debug your refrigerator?) They also depend on countless nonlinear network effects, sleeper effects, and other nuisances. The popularity of baby names (Mildred, Deborah, Jennifer, Chloe), and the rates of homicide (down in the 1940s, up in the 1960s, down again in the 1990s) are just two of the social trends that fluctuate wildly in defiance of the best efforts of social scientists to explain them after the fact, let alone predict them beforehand.

But if you insist. The new millennium saw the introduction of direct-to-consumer genomics. A number of new companies have been recently launched. You can get everything from a complete sequencing of your genome (for a cool $350,000), to a screen for more than a hundred Mendelian disease genes, to a list of traits, disease risks, and ancestry data. Here are some possible outcomes:

- Personalized medicine, in which drugs are prescribed according to the patient's molecular background rather than by trial and error, and in which prevention and screening recommendations are narrow-casted to those who would most benefit.
- An end to many genetic diseases. Just as Tay-Sachs has been almost wiped out in the decades since Ashkenazi Jews have tested themselves for the gene, a universal-carrier screen, combined with preimplantation genetic diagnosis for carrier couples who want biological children, will eliminate a hundred others.

- Universal insurance for health, disability, and home care. Forget the political debates about the socialization of medicine. Cafeteria insurance will no longer be actuarially viable if the highest-risk consumers can load up on generous policies while the low-risk ones get by with the bare minimum.
- An end to the genophobia of many academics and pundits, whose blank-slate doctrines will look increasingly implausible as people learn about genes that affect their temperament and cognition.
- The ultimate empowerment of medical consumers, who will know their own disease risks and seek commensurate treatment, rather than relying on the hunches and folklore of a paternalistic family doctor.

But then again, maybe not.

OUR GENES ARE NOT OUR FATE

—◇—

DEAN ORNISH

DEAN ORNISH is a medical doctor, clinical professor of medicine at the University of California, San Francisco, and the author of *The Spectrum: A Scientifically Proven Program to Feel Better, Live Longer, Lose Weight, and Gain Health*.

We are entering a new era of personalized medicine. One size does not fit all.

One way to change your genes is to make new ones, as Craig Venter has elegantly shown. Another is to change your lifestyle: what you eat, how you respond to emotional stress, whether or not you smoke cigarettes, how much you exercise, and the experience of love and intimacy.

New studies show that these comprehensive lifestyle changes may change gene expression in hundreds of genes in only a few months—"turning on" (up-regulating) disease-preventing genes and "turning off" (down-regulating) genes that promote heart disease, oncogenes that promote breast cancer and prostate cancer, and genes that promote inflammation and oxidative stress. These lifestyle changes also increase telomerase, the enzyme that repairs and lengthens telomeres—the ends of our chromosomes, which control how long we live.

As genomic information for individuals becomes more widely available—via the decoding of each person's complete genome (as Venter and James Watson have done) or partially (and less expensively) via new personal genomics companies—this information will be a powerful motivator for people to make comprehensive lifestyle changes that may beneficially affect their gene expression and significantly reduce the incidence of the pandemic of chronic diseases.

A FOREBRAIN FOR THE WORLD MIND

―◁◦▷―

W. DANIEL HILLIS

W. DANIEL HILLIS is a physicist and computer scientist, chairman of Applied Minds, Inc., and the author of *The Pattern on the Stone: The Simple Ideas That Make Computers Work.*

In 1851, Nathaniel Hawthorne wrote, "Is it a fact—or have I dreamt it—that, by means of electricity, the world of matter has become a great nerve, vibrating thousands of miles in a breathless point of time? Rather, the round globe is a vast head, a brain, instinct with intelligence!" He was writing about the telegraph, but today we make essentially the same observation about the Internet.

One might suppose that with all its zillions of transistors and billions of human minds, the world brain would be thinking some pretty profound thoughts. There is little evidence that this is so. Today's Internet functions mostly as a giant communications-and-storage system, accessed by individual humans. Although much of human knowledge is represented in some form within the machine, it is not yet represented in a form that is particularly meaningful to the machine. For the most part, the Internet knows no more about the information it handles than the telephone system knows about the conversations that take place over its lines. Most of those zillions of transistors are either doing something very trivial or nothing at all, and most of those billions of human minds are doing their own thing.

If there is such a thing as a world mind today, then its thoughts are primarily about commerce. It is the "invisible hand" of Adam Smith, deciding the prices, allocating the capital. Its brain is composed not only of the human buyers and sellers but also of the trading programs on Wall Street and the economic models of the central banks. The wires "vibrating thousands of miles in a breathless point of time" are not just carrying messages between human minds; they are participating in the decisions of the world mind as a whole. This unconscious system is the world's hindbrain.

I call this the hindbrain because it is performing unconscious functions necessary to the organism's survival—functions so primitive that they predate development of the brain. Included in this hindbrain are the functions of preference and attention that create celebrity, popularity, and fashion, all fundamental to the operation of human society. This hindbrain is ancient. Although it has been supercharged by technology, growing in speed and capacity, it has grown little in sophistication. This global hindbrain is subject to mood swings and misjudgments, leading to economic depressions, panics, witch hunts, and fads. It can be influenced by propaganda and by advertising. It is easily misled. As vital as the hindbrain is for survival, it is not very bright.

What the world mind really needs is a forebrain with conscious goals, access to explicit knowledge, and the ability to reason and plan. A world forebrain would need the ability to perceive collectively, to decide collectively, and to act collectively. Of these three functions, our ability to act collectively is the most developed.

For thousands of years, we have understood methods for breaking a goal into subgoals that can be accomplished by separate teams, and for recursively breaking the subgoals down again and again, until they can be accomplished by individuals. This management by hierarchy scales well. I can imagine that the construction of the pyramids was a celebration of its discovery. The hierarchical teams that built those monuments were an extension of the pharaoh's body, the pyramid a dramatic demonstration of his power to coordinate the efforts of many. Pyramid builders had to keep their direct reports within shouting distance, but electronic communication has allowed us to extend our virtual bodies—literally, corporations—to a global scale. The Internet has even allowed such composite action to organize itself around an established goal, without the pharaoh. The Wikipedia is our Great Pyramid.

The collective perception of the world mind is also relatively well developed. The most important recent innovations have been search and recommendation engines, which combine the inputs of humans with machine algorithms to produce a useful result. This is another area where scale helps. Many eyes and many judgments are combined into a collective perception that is beyond the scope of any individual. The weak point is that the result of all this collective perception is just a recommendation list. For the world mind to truly perceive, it will need a way of sharing more general forms of knowledge, in a format that can be understood by both humans and machines. Various new companies are beginning to do just that.

What is still missing is the ability of a group of people (or people and machines) to make collective decisions with intelligence greater than the individual. This can sometimes be accomplished in small groups through conversation, but the method does not scale well. Generally speaking, technology has made the conversation larger but not smarter. For large groups, the state-of-the-art method for collective decision making is still the vote. Voting only works to the degree that, on average, each voter is able to individually determine the right decision. This is not good enough. We need an intelligence that will scale with the size of our problems.

So this is the development that will make a difference: a

method for groups of people and machines to work together to make decisions in a way that takes advantage of scale. With such a scalable method for collective decision making, our zillions of transistors and billions of brains can be used to advantage, giving the collective mind a way to focus our collective actions. Given this, we will finally have access to intelligence greater than our own. The world mind will finally have a forebrain, and this will change everything.

FUTURE AS PRESENT: A FINAL EXPERIMENT

◄◦►

ERNST PÖPPEL

ERNST PÖPPEL is a neuroscientist. He is chairman of the Human Science Center of Munich University and director of the Parmenides Center of the Study of Thinking in Munich and the author of *Mindworks: Time and Conscious Experience*.

When time came to an end, the gods decided to run a final experiment. They wanted to be prepared after the Big Crunch for potential trajectories of life after the next Big Bang. For their experiment, they choose two planets in the universe where evolution had resulted in similar developments of life. For planet 1, they decided to interfere with evolution by allowing only one species to develop its brain to a high level of complexity. This species referred to itself as "intelligent." Members of this species were proud of their achievements in science, technology, the arts, and philosophy.

For planet 2, the gods altered just one variable. For this planet, they allowed *two* species with high intelligence to develop. The two species shared the environment, but—and this was crucial to the divine experiment—they did not communicate directly with each other. Direct communication was limited to their own species only. Thus, one species could not directly inform the other about its plans; each species could register only what had happened to its common environment.

The question was: How would life be managed on planet 1 and planet 2? As for any organism, the goal on both planets was to maintain an internal balance or homeostasis by optimal use of the available resources. As long as the members of the different samples were not too intelligent, stability was maintained; however, when they became more intelligent and, according to their own view, really smart, and when the frame of judgment changed—that is, individual interests became dominant—trouble was preprogrammed. Because of uncontrolled personal greed, more resources were drawn from the environment than could be replaced. Which planet would do better, with such species of too much intelligence, to maintain the conditions of life?

Data analysis after the experimental period of two hundred years showed that planet 2 was much better at maintaining the stability of the environment. Why? The species on planet 2 had always to monitor the consequences of actions of the other species. If one species took too many resources for individual satisfaction, sanctions by the other species would be the consequence. Thus, drawing resources from the environment was controlled by the other species in a bidirectional way resulting in a dynamic equilibrium.

When the gods published their results, they drew the following conclusions: Long-term stability in complex systems—such as social systems with members of too much intelligence—can be maintained if two complementary systems interact with each other. If only one system has developed, as on planet 1, the gods recommended adoption of a second system for regulative purposes. For social systems, that should be the next generation, whose future environment should be made present both conceptually and emotionally. Thus, long-term stability will be guaranteed.

Being good brain scientists, the gods knew that making the future present is not simply a matter of abstract or explicit knowledge, which is necessary but not sufficient for action resulting in a long-term equilibrium. Decisions have to be anchored in the emotions as well; an empathic relationship between the members of the two systems has to be developed. If the future becomes present, it can in the future be a present.

BUT WE SHALL ALL BE CHANGED

———◁◦▷———

FRANK J. TIPLER

FRANK J. TIPLER is a professor of mathematical physics at Tulane University and the author of *The Physics of Immortality: Modern Cosmology, God, and the Resurrection of the Dead* and *The Physics of Christianity*.

I'm sixty-two, so I'll have to limit my projections to what I expect to happen in the next two to three decades. I believe these will be the most interesting times in human history. (Remember the old Chinese curse about "interesting times"?) Humanity will see, before I die, the Singularity—the day when we finally create a human-level artificial intelligence. This involves considering the physics advances that will be required to create the computer capable of running a strong AI program.

By both my calculations and those of Ray Kurzweil, the originator of the Singularity idea, the 10-teraflop speed of today's supercomputers means they have more than enough computing power to run a minimum AI program, but we are missing some crucial idea in this program. The British mathematician John Conway's Game of Life has been proven to be a universal program capable of expressing a strong AI program, and it should therefore be able (if allowed to run long enough) to bootstrap itself into the complexity of human-level intelligence. But Game of Life programs do not do this. They increase their complexity just so far and then stop—why, we don't know. As I said, we are missing something, and what we are missing is the key to human creativity.

But an AI program can be generated by brute force. We can map an entire human personality, together with a simulated environment, into a program and run it. Such a program would be roughly equivalent to the program being run in the movie *The Matrix*, and it would require enormous computing power, far beyond that of today's supercomputers. The power required can be provided only by a quantum computer.

A quantum computer works by parallel processing across the multiverse. That is, part of the computation is done in this universe by you and your part of the quantum computer, and the other parts of the computation are done by your analogs with their parts of the computer in the other universes of the multiverse. The full potential of the quantum computer has not been realized, because the existence of the multiverse has not yet been accepted—even by workers in the field of quantum computation—in spite of the fact that the multiverse's existence is required by quantum mechanics (and by classical mechanics in its most powerful form, Hamilton-Jacobi theory).

Other new technologies become possible via action across the multiverse. For example, the Standard Model of particle physics, the theory of all forces and particles except gravity, confirmed by many experiments in the past forty years, tells us that it is possible to transcend the laws of conservation of baryon number (the number of protons plus neutrons) and conservation of lepton number (the number of electrons plus neutrinos) and thereby convert matter into energy in a process far more efficient than nuclear fission or fusion. According to the Standard Model, the proton and electron making up a hydrogen atom can be combined to yield pure energy in the form of photons, or neutrino-antineutrino pairs. If the former, then we would have a mechanism that would allow us to convert garbage into energy, a device that Doc, in the movie *Back to the Future*, obtained from his trip to the future. If the latter, then the directed neutrino-antineutrino beam would provide the ultimate rocket; the exhaust would be invisible—just like that of the propulsion mechanism Doc also borrowed from the future. The screenwriters got it right: Doc's devices are indeed in the offing.

A quantum computer running an AI program, direct conversion of matter into energy, the ultimate rocket enabling the AIs and the human downloads to begin interstellar travel at near-light-speed—all these depend on the same physics and should appear at the same time in the future. Provided we have the courage to develop the technology, allowed by the known laws of physics. I have grave doubts that we will.

In order to have advances in physics and engineering, one must first have physicists and engineers. The number of students majoring in these subjects has dropped enormously in the quarter century that I have been a professor. Worse, the quality of the students we do have has dropped precipitously. The next decade will see the retirement of Stephen Hawking and others less well known but of similar ability. I know of no one of remotely equal creativity to replace them. Small wonder, given that the starting salary of a Wall Street lawyer fresh out of school is currently three times my own physicist's salary. As a result, most American engineers and physicists are now foreign-born.

But can foreign countries continue to supply engineers and physicists? That is, will engineers and physicists be available in *any* country? The birthrate of most developed nations has been far below replacement level for a decade and more. This birth dearth also holds for China, due to their one-child policy and, remarkably, is developing even in the Muslim and the southern nations. We may not have enough people in the next twenty years to sustain the technology we already have, to say nothing of developing the one I've just described.

The great Galileo scholar Giorgio De Santillana, who taught me history of science when I was an undergraduate at MIT in the late 1960s, wrote that Greek scientific development ended in the century or so before the Christian era because of a birth dearth and a simultaneous bureaucratization of intellectual inquiry. I fear we are seeing a repeat of this historical catastrophe today. However, I remain cautiously optimistic that we will develop the ultimate technology described above and transfer it, with faltering hands, to our ultimate successors, the AIs and the human downloads, who will be thus enabled to expand outward into interstellar space, engulf the universe, and live forever.

THE CREDIT CRUNCH FOR MATERIALISM

—◦—

RUPERT SHELDRAKE

RUPERT SHELDRAKE is director of the Perrott-Warrick Project for research on unexplained human and animal abilities, Trinity College, Cambridge, and the author of *A New Science of Life: The Hypothesis of Formative Causation*.

Credit crunches happen because of too much credit and too many bad debts. Credit is literally belief, from the Latin *credo*, "I believe." Once confidence ebbs, the loss of trust is self-reinforcing. The game changes.

Something similar is happening with materialism. Since the nineteenth century, its advocates have promised that science will explain everything in terms of physics and chemistry; science will show that there is no God and no purpose in the universe; it will reveal that God is a delusion inside human minds and hence in human brains; and it will prove that brains are nothing but complex machines.

Materialists are sustained by the faith that science will redeem their promises, turning their beliefs into facts. Meanwhile, they live on credit. The philosopher of science Sir Karl Popper described this faith as "promissory materialism," because it depends on promissory notes for discoveries not yet made. Despite all the achievements of science and technology, they face an unprecedented credit crunch.

In 1963, when I was studying biochemistry at Cambridge, I

was invited to a series of private meetings with Francis Crick and Sydney Brenner in Brenner's rooms in King's College, along with a few of my classmates. Crick and Brenner had recently cracked the genetic code. Both were ardent materialists. They explained that there were two major unsolved problems in biology: development and consciousness. These had not been solved because the people who worked on them were not molecular biologists—nor very bright. Crick and Brenner were going to find the answers within ten years, or maybe twenty. Brenner would take development, and Crick consciousness. They invited us to join them.

Both tried their best. Brenner was awarded the Nobel Prize in 2002 for his work on the development of the nematode worm *Caenorhabditis elegans*. Crick corrected the manuscript of his final paper on the brain the day before he died, in 2004. At his funeral, his son, Michael, said that what made him tick was not the desire to be famous, wealthy, or popular, but "to knock the final nail into the coffin of vitalism."

He failed. So did Brenner. The problems of development and consciousness remain unsolved. Many details have been discovered, dozens of genomes have been sequenced, and brain scans are ever more precise. But there is still no proof that life and minds can be explained by physics and chemistry alone.

The fundamental proposition of materialism is that matter is the only reality. Therefore consciousness is nothing but brain activity. However, among researchers in neuroscience and consciousness studies, there is no consensus. Leading journals such as *Behavioral and Brain Sciences* and the *Journal of Consciousness Studies* publish many articles that reveal deep problems with the materialist doctrine. For example, Steven Lehar argues that inside our heads there must be a miniaturized virtual-reality full-color three-dimensional replica of the world. When we look at the sky, the sky is in our heads; our skulls are beyond the sky. Others, like the psychologist Max Velmans, argue that virtual-reality displays are not confined to our brains; they are life size, not miniaturized. Our visual perceptions are outside our skulls, just where they seem to be.

The philosopher David Chalmers has called the very existence of subjective experience the "hard problem" of consciousness, because it defies explanation in terms of mechanisms. Even if we understand how eyes and brains respond to red light, for example, the quality of redness is still unaccounted for.

In biology and psychology, the credit rating of materialism is falling fast. Can physics inject new capital? Some materialists prefer to call themselves physicalists, to emphasize that their hopes depend on modern physics, not nineteenth-century theories of matter. But physicalism's credit rating has been reduced by physics itself, for four reasons:

First, some physicists argue that quantum mechanics cannot be formulated without taking into account the minds of observers; hence minds cannot be reduced to physics, because physics presupposes minds.

Second, the most ambitious unified theories of physical reality—superstring and M theory, with ten and eleven dimensions, respectively—take science into completely new territory. They are a very shaky foundation for materialism, physicalism, or any other preestablished belief system. They are pointing somewhere new.

Third, the known kinds of matter and energy constitute only about 4 percent of the universe. The rest consists of dark matter and dark energy. The nature of 96 percent of reality is literally obscure.

Fourth, the cosmological anthropic principle asserts that if the laws and constants of nature had been slightly different at the moment of the Big Bang, biological life could never have emerged and hence we would not be here to think about it. So did a divine mind fine-tune the laws and constants in the beginning? Some cosmologists prefer to believe that our universe is one of a vast, and perhaps infinite, number of parallel universes, all with different laws and constants. We just happen to exist in the one that has the right conditions for us.

In the eyes of skeptics, the multiverse theory is the ultimate violation of Ockham's razor, the principle that entities should not be multiplied unnecessarily. But even so, it does not succeed in getting rid of God. An infinite God could be the God of an infinite number of universes.

Here on Earth, we are facing climate change, great economic uncertainty, and cuts in science funding. Confidence in materialism is draining away. Its leaders, like central bankers, keep printing promissory notes, but it has lost its credibility as the central dogma of science. Many scientists no longer want to be 100 percent invested in it.

Materialism's credit crunch changes everything. As science is liberated from this nineteenth-century ideology, new perspectives and possibilities will open up—not just for science, but for other areas of our culture that are dominated by materialism. And by giving up the pretense that the ultimate answers are already known, the sciences will be freer—and more fun.

THE LAPTOP QUANTUM COMPUTER

———◆◇◆———

DONALD D. HOFFMAN

DONALD D. HOFFMAN is a cognitive scientist at the University of California at Irvine and the author of *Visual Intelligence: How We Create What We See.*

Everything will change with the advent of the laptop quantum computer. The transition from PCs to QCs will not merely continue the doubling of computing power in accord with Moore's Law. It will induce a paradigm shift, both in the power of computing (at least for certain problems) and in the conceptual frameworks we use to understand computation, intelligence, neuroscience, social interactions, and sensory perception.

Today's PCs depend, of course, on quantum mechanics for their proper operation. But their computations do not exploit two computational resources unique to quantum theory: superposition and entanglement. To call these "computational resources" is already a major conceptual shift. Until recently, superposition and entanglement were regarded primarily as mathematically well-defined but psychologically incomprehensible oddities of the quantum world—fodder for interminable and apparently unfruitful philosophical debate. But they turn out to be more than idle curiosities. They are bona fide computational resources that can solve certain problems that are intractable using classical computers. The best-known example is Peter Shor's quantum-algorithm, which can, in principle, break encryptions impene-trable to classical algorithms.

The issue is the "in principle" part. Quantum theory is well established, and quantum computation, although a relatively young discipline, has an impressive array of algorithms that can, in principle, run circles around classical algorithms on several important problems. But what about in practice? Not yet, and not by a long shot. There are formidable materials—science problems that must be solved—such as instantiating quantum bits (qubits) and quantum gates and avoiding an unwanted noise called decoherence—before the promise of quantum computa-tion can be fulfilled by tangible quantum computers. Many ex-perts bet that the problems can't adequately be solved. I think this bet is premature. We will have laptop QCs, and they will transform our world.

When laptop QCs become commonplace, they will natu-rally lead us to rethink the notion of intelligence. At present, in-telligence is modeled by computations (sometimes simple and sometimes complex) that allow a system to learn—often by in-teracting with its environment—how to plan, reason, generalize, and act to achieve goals. The computations might be serial or parallel, but they have heretofore been taken to be classical.

One hallmark of a classical computation is that it can be traced—that is, one can in principle observe the states of all the variables at each step of the computation. This is helpful for debugging. But one hallmark of quantum computations is that they cannot, in general, be traced. Once the qubits have been initialized and the computation started, you cannot observe in-termediate stages of the computation without destroying it. You aren't allowed to peek at a quantum computation while it's in progress. The full horsepower of a quantum computation is un-leashed only when, so to speak, you don't look. This is jarring. It clashes with our classical way of thinking about computation. It also clashes with our classical notion of intelligence. In the quantum realm, intelligence happens when you don't look. Insist on looking and you destroy this intelligence. We will be forced to reconsider what we mean by "intelligence" in light of quantum computation. In the process, we might find new conceptual tools for understanding those creative insights that seem to come from the blue—that is, whose origin and development can't seem to be traced.

Laptop QCs will make us rethink neuroscience, too. A few decades ago, we peered inside brains and saw complex telephone switchboards. Now we peer inside brains and see complex classical computations, both serial and parallel. What will we see once we have thoroughly absorbed the mindset of quantum computation? Some say we will still find only classical computations, because the brain and its neurons are too massive for quantum effects to survive. But evolution by natural selection leads to surprising adaptations, and there might in fact be selective pressures toward quantum computations.

One case in point arises in a classic problem of social interaction: the Prisoner's Dilemma. In one version of this dilemma, someone yells "Fire!" in a crowded theater. Each person in the crowd has a choice. They can cooperate with everyone else, by exiting in turn, in an orderly fashion. Or they can defect and bolt for the exit. Everyone cooperating would be best for the whole crowd; it is a so-called Pareto optimal solution. But defecting is best for each individual; it is a so-called Nash equilibrium.

What happens is that everyone defects, and the crowd as a whole suffers. But this problem of the Prisoner's Dilemma— namely, that the Nash equilibrium is not Pareto optimal—is an artifact of the classical computational approach to the dilemma. There are quantum strategies, involving superpositions of cooperation and defection, for which the Nash equilibrium is Pareto optimal. In other words, the Prisoner's Dilemma can be resolved, and the crowd as a whole needn't suffer, if quantum strategies are available. If the Prisoner's Dilemma is played out in an evolutionary context, there are quantum strategies that drive all classical strategies to extinction. This is suggestive. Could there be selective pressures that built quantum strategies into our nervous systems and our social interactions? Do such strategies provide a way to rethink the notion of altruism—perhaps as a superposition of cooperation and defection?

Laptop QCs will also alter our view of sensory perception. Superposition seems to be telling us that our sensory representations, which carve the world into discrete objects with properties such as position and momentum, are an inadequate description of reality: No definite position or momentum can be ascribed to, say, an electron when it is not being observed. Entanglement seems to be telling us that the very act of carving the world into discrete objects is an inadequate description of reality: Two electrons, billions of light-years apart in our sensory representations, are in fact intimately and instantly linked as a single entity.

When superposition and entanglement cease to be abstract curiosities and become computational resources indispensable to the function of our laptops, they will transform our understanding of perception and of the relation between perception and reality.

UNDO THE PRESENT; RECALL THE PAST

SETH LLOYD

SETH LLOYD is a quantum-mechanical engineer at MIT and the author of *Programming the Universe: a Quantum Computer Scientist Takes on the Cosmos.*

My job is to design and build quantum computers, computers that store and process information at the level of individual atoms. Even at the current rapid rates of progress in current computer technology, with the computer components halving in size every two years or less and computers doubling in power over the same time, quantum computers should not be available for forty years. Yet we are building simple quantum computers today. I could tell you that quantum computers will drastically change the way the world works during our lifetime. But I'm not going to do that, for the simple reason that I have no idea whether it's true or not.

Whether or not they change the world, quantum computers have something to offer to all of us. When they flip those atomic bits to perform their computations, quantum computers possess several useful features. It's well known that quantum computers, properly programmed, afford their users privacy and anonymity guaranteed by the laws of physics. A less well-known virtue of quantum computers is that everything that they do, they can undo as well. This ability is built into quantum computers at the level of fundamental physical law. At their most microscopic level, the laws of physics are reversible: What goes forward can go backward. (By contrast, at the more macroscopic scales at which classical computers operate, the second law of thermodynamics kicks in, and what is done cannot be undone.) Because they operate at the level of individual atoms, quantum computers inherit those atoms' ability to undo the present and recall the past.

While quantum computers afford their users protection and anonymity that classical computers cannot, even classical computers can be programmed to share this ability to erase regret, although they currently are not. Although classical computers dissipate heat and operate in a physically irreversible way, they can still function in a logically reversible fashion. Properly programmed, they can unperform any computation they can perform. We already see a hint of this digital nostalgia in hard-disk "time machines," which restore a disk to its state in an earlier, precrash era.

Suppose we were to put this ability of computers to run the clock backward to the service of undoing not merely our accidental erasures and unfortunate viral infections but also financial transactions conducted under fraudulent conditions? Credit card companies already supply us with protection against theft conducted in our name. Why shouldn't more important financial transactions be similarly guaranteed? Contracts for home sales, stock deals, and credit default swaps are recorded and executed digitally. What would happen if we combined digital finance with reversible computation?

For example, if a logically reversible computer—quantum or classical—were used to record a financial contract and execute its terms, then at some later point, if the parties were not satisfied with the way those terms were executed, they could be unexecuted—any money disbursed could be reimbursed and the contract could be deleted as if it had never been. Since finance is already digital, why not introduce a digital time machine? Let's agree now that when the next crash comes we'll restore everything to a better, earlier time, before we clicked those inauspicious buttons and brought on the blue screen of financial death.

Can it be done? The laws of physics and computation say yes. But what about the laws of human nature? The financial time machine erases profits as well as losses. Will hedge-fund managers and Ponzi schemers sign on to turn back the clock if schemes go awry, even if it means that their gains, well gotten or ill, will be restored to their clients/victims? If they refuse to agree, then you don't have to give them your money.

I make no predictions, but the laws of physics have been around for a long time. Meanwhile, the only true "law" of human nature is its intrinsic adaptability. Microscopic reversibility is the way that nature does business. Maybe we can learn a thing or two from nature to change the way *we* do business.

ROUNDING AN ENDLESS VICIOUS CIRCLE

———◦———

ALAN ALDA

ALAN ALDA is an actor, writer, and director, and the host of the PBS program *Scientific American Frontiers*.

I find it hard to believe that *anything* will change everything. The only exception might be if we suddenly learned how to live with one another. But does anyone think that will come about in a foreseeable lifetime?

Evidence from the past seems to point to our becoming increasingly dangerous nearly every time we come up with a new idea or technology. These new things are usually wholesome and benign at first (movable type, pharmacology, rule of law), but before long we find ways to use these inventions to do what we do best: exercise power over one another.

Even if we were visited by weird little people from another planet and were forced to band together, I suspect we'd soon be finding ways to break into factions again, identifying those among us who are not quite people.

We keep rounding an endless vicious circle. Will an idea or technology emerge anytime soon that will let us exit this lethal cyclotron before we meet our fate head-on and scatter into a million pieces? Will we outsmart our own brilliance before this planet is painted over with yet another layer of people? Maybe, but I doubt it.

THE IDEA OF NEGATIVE AND IATROGENIC SCIENCE

NASSIM NICHOLAS TALEB

NASSIM NICHOLAS TALEB is an epistemologist of randomness, an applied statistician, and the author of *The Black Swan: The Impact of the Highly Improbable*.

People want advice on how to get rich—and they'll pay for it. Now, how not to go bust does not appear to be valid advice, yet given that over time only a minority of companies do not go bust, avoiding death is the best possible—and most robust—advice. It is particularly good advice after your competitors get in trouble and you can legally pillage their businesses. But few value such advice: This is the reason Wall Street quants, consultants, and investment managers are in business, in spite of their charlatanic record. I was recently on TV and some empty suit kept bugging me for precise advice on how to pull out of the present crisis. It was impossible to communicate my "what not to do" advice—or that my field is error avoidance, not emergency-room surgery, and that it is a stand-alone discipline. Indeed, I have spent twelve years trying to explain that in many instances no models were better—and wiser—than the mathematical acrobatics we performed in finance, and that it took a monumental crisis to convince people of the point.

Unfortunately, such lack of rigor pervades the place where we expect it the least: institutional science. Science, particularly its academic version, never liked negative results, let alone the statement and advertising of its own limits; the reward system is not set up for it. You get respect for doing funambulism or spectator sports—following the right steps to become the "Einstein of economics" or the "next Darwin," rather than giving society something real by debunking myths or marking where our knowledge stops. In some instances, we accept a limit to knowledge—trumpeting, say, Gödel's "breakthrough" incompleteness theorems—because it shows elegance in formulation and mathematical prowess, though the importance of such a limit is dwarfed by our practical limits in forecasting climate changes, crises, social turmoil, or the fate of the endowment funds that will finance research of such future elegance.

Let's consider medicine, which started saving lives only less than a century ago (I am generous) and to a lesser extent than initially advertised in the popular literature, as the declining mortality rate seems to arise more from improvements in sanitation and the (random) discovery of antibiotics than from therapeutic contributions. Doctors, driven by the beastly illusion of control, spent a long time killing patients instead of considering that "doing nothing" could be a valid option—and research compiled by my colleague Spyros Makridakis shows that they still do, to some extent. Indeed, practitioners who were conservative and considered simply letting nature do its job, or stated the limit of our medical understanding, were until the 1960s accused of "therapeutic nihilism." It was deemed "unscientific" to decide on a course of action based on an incomplete understanding to the human body—to say, "This is the limit where my body of knowledge stops."

The very term "iatrogenic"—that is, harm caused by the healer—is not ubiquitous. I have never seen it used outside of medicine. In spite of my lifelong obsession with what is called "type-two error," or false positive, I was introduced to the concept only very recently, thanks to a conversation with the essayist Bryan Appleyard. How can such a major idea remained hidden from our consciousness? Even in medicine—that is, modern medicine—the ancient concept of "Do no harm" sneaked in very late. The philosopher of science Georges Canguilhem wondered why it was not until the 1950s that the idea came to us. This, to me, is a mystery: how professionals can cause harm for such a long time in the name of knowledge and get away with it.

Sadly, further investigation shows that these iatrogenics were mere rediscoveries after science got too arrogant in the Enlightenment. Alas, once again, the elders knew better: Greeks, Romans, Byzantines, and Arabs had a built-in respect for the limits of knowledge. There is a treatise by the medieval Arab philosopher and doctor Al-Ruhawi which demonstrates the familiarity of these Mediterranean cultures with iatrogenics. I have also speculated that religion saved lives by taking the patient away from the doctor. You could satisfy your illusion of control by going to the Temple of Apollo rather than by seeing the doctor. What is interesting is that the ancient Mediterraneans may have understood the trade-off very well and have accepted religion partly as a tool to tame such illusion of control.

You cannot do anything with knowledge unless you know where it stops and the costs of using it. Post-Enlightenment science and its daughter, superstar science, were lucky to have done well in (linear) physics, chemistry, and engineering. But at some point, we need to give up on elegance to focus on something that has been given short shrift for a very long time: the maps showing what current knowledge and current methods do not do for us. We need a rigorous study of generalized scientific iatrogenics, of what harm can be caused by science—or, better, an exposition of what harm has been done by science. I find it the most respectable of pursuits.

THE FEELING THAT THINGS WILL GET WORSE

◆

BRIAN ENO

BRIAN ENO is an artist, musician, composer, and record producer, including albums by U2, Talking Heads, and Coldplay.

What would change everything is not even a thought. It's more of a feeling.

Human development thus far has been fueled and guided by the feeling that things could be, and probably will be, better. The world was rich compared to its human population; there were new lands to conquer, new thoughts to nurture, and new resources to fuel it all. The great migrations of human history grew from the feeling that there was a better place, and the institutions of civilization grew out of the feeling that checks on pure individual selfishness would produce a better world for everyone involved, in the long term.

What if this feeling changes? What if we come to feel as though there were no "long term"—or not one to look forward to? What if, instead of feeling that we are standing at the edge of a wild new continent full of promise and hazard, we started to feel that we're on an overcrowded lifeboat in hostile waters, fighting to stay on board, prepared to kill for the last scraps of food and water?

Many of us grew up among the reverberations of the 1960s. At that time, there was a sense that the world could be a better place and that our responsibility was to make it real by living it. Why did this feeling take root? Probably because there was new wealth around, a new unifying mass culture, and a newly empowered generation whose life experience told it that the line on the graph could point only up. In many ways, this idealism paid off: The better results remain with us today—surfacing, for example, in the wiki-ized world of idea sharing, of which this conversation is a part.

But suppose the feeling changes. Suppose that people start to anticipate the future world not in that way but instead as something more closely resembling the nightmare of desperation, fear, and suspicion described in Cormac McCarthy's post-cataclysm novel *The Road*. What happens then?

The following: Humans fragment into tighter, more selfish bands. Big institutions, because they operate on long timescales and require structures of social trust, don't cohere; there isn't time for them. Long-term projects are abandoned; their payoffs are too remote. Global projects are abandoned—not enough trust to make them work. Resources that are already scarce will be rapidly exhausted, as everybody tries to grab the last precious bits. Any kind of social or global mobility is seen as a threat and harshly resisted. Freeloaders and brigands and pirates and cheats take control. Survivalism rules. Might makes right.

This is a dark thought, but one to keep an eye on. Feelings are more dangerous than ideas, because they aren't susceptible to rational evaluation. They grow quietly, spreading underground, and erupt suddenly, all over the place. They can take hold quickly and run out of control (*"Fire!"*), and by their nature they tend to be self-fueling. If our world becomes gripped by this particular feeling, everything it presupposes could soon become true.

HOMESTEADING IN HILBERT SPACE

FRANK WILCZEK

FRANK WILCZEK is a physicist at MIT, a Nobel laureate, and the author of *The Lightness of Being: Mass, Ether, and the Unification of Forces*.

More than a hundred years passed between Columbus's first, confused sighting of America in 1492 and the vanguard of English colonization at Jamestown in 1607. A similar interval separates us today from Max Planck's first confused sighting of the quantum world in 1899. The quantum world is a new New World, far more alien and difficult of access than Columbus's old New World. It is also, in a real sense, much bigger. While discovery of the old New World roughly doubled the land area available to humans, the new New World exponentially expands the *dimension* of physical reality. (This is established physics, independent of speculations about extra spatial dimensions, which are essentially classical.) For example, every single electron's spin doubles that dimension. Our fundamental equations do not live in the three-dimensional space of classical physics but in an effectively infinite-dimensional space: Hilbert space. It will take us much more than a century to homestead that new New World, even at today's much accelerated pace.

We have managed to establish some beachheads, but the vast interior remains virgin territory, unexploited. (This time, presumably, there are no aboriginals.) Poking along the coast, we have already stumbled upon transistors, lasers, superconducting magnets, and a host of other gadgets. What's next? I don't know for sure, of course, but there are two everything-changers that seem safe bets:

- New microelectronic information processors, informed by quantum principles—perhaps based on manipulating electron spins or on supplementing today's silicon with graphene—will enable more cycles of Moore's Law on several fronts: smaller, faster, cooler, cheaper. Supercomputers will approach and then surpass the exaflop frontier, making their capacity comparable to that of human brains. Improved bandwidth will put the Internet on steroids, allowing instant access from anywhere to all the world's information and blurring or obliterating the experienced distinction between virtual and physical reality.

- Designer materials better able to convert energy from the hot and unwieldy quanta (photons) that the sun rains upon us into more convenient forms (chemical bonds) will power a new economy of abundance. Evolution, in its patient blindness, managed to develop photosynthesis; with mindful insight, we will do better.

As we thus augment our intelligence and our power, a sort of bootstrap may well come into play. We—or our machines, or our hybrid descendants—will acquire the wit and strength to design and construct still better minds and engines, in an ascending spiral.

Our creative mastery over matter through quantum theory is still embryonic. The best is yet to come.

REVELATION

———◦———

STEFANO BOERI

STEFANO BOERI is an architect at the Politecnico of Milan and the editor of *Abitare*.

What will change everything? Discovering that someone from the future has already come to visit us.

THE DISCOVERY OF INTELLIGENT LIFE FROM SOMEWHERE ELSE

———◇———

DOUGLAS RUSHKOFF

DOUGLAS RUSHKOFF is a media analyst, documentary writer, and the author of *Life Inc.: How the World Became a Corporation and How to Take It Back.*

We're talking about changing *everything*—not just our abilities, relationships, politics, economy, religion, biology, language, mathematics, history, or future but all of these things at once. The only single event I can see shifting pretty much everything at once is our first encounter with intelligent extraterrestrial life.

The development of any of our current capabilities—genetics, computing, language, even compassion—all feel like incremental advances in existing abilities. As we've seen before, the culmination of one branch of inquiry always just opens the door to a new branch and never yields the wholesale change of state we anticipated. Nothing we've done in the past couple of hundred thousand years has truly changed everything, so I don't see us doing anything in the future that would change everything, either.

No, the only way to change everything is for something to be done *to* us instead. The encounter between humanity and an "other" entails a shift beyond the solipsism that has characterized our civilization since our civilization was born. It augurs a reversal as big as the encounter between an individual and its offspring or a creature and its creator. Even if this were to be the result of something we'd done, it would now be independent of us and our efforts.

To meet a neighbor, whether outer, inner, cyber-, or hyperspatial, finally turns us into an "us." To encounter an other—whether god, ghost, biological sibling, independently evolved life-form, or emergent intelligence of our own creation—changes what it means to be human.

Our computers may never inform us that they are self-aware; extraterrestrials may never broadcast a signal to our SETI dishes; interdimensional creatures may never appear to those who aren't taking psychedelics at the time—but if any of these things happened, that would change everything.

A CURE FOR HUMANKIND'S EXISTENTIAL LONELINESS

—◅◦▻—

PAUL SAFFO

PAUL SAFFO is a technology forecaster.

Accelerating change is the new normal. Even the most dramatic discoveries waiting in the wings will do little more than push us further along the rollercoaster of exponential arcs that define modern life. Momentous discoveries compete with Hollywood gossip for headline space, as a public accustomed to a steady diet of surprises reacts to the latest astonishing science news with a shrug.

But there is one development that would fundamentally change everything: the discovery of nonhuman intelligences equal or superior to our species. It would change everything because our crowded, quarreling species is lonely. Vastly, achingly, existentially lonely. It is what compels our faith in gods whose existence lies beyond logic or proof. It is what animates our belief in spirits and fairies, ghosts, and little green men. It is why we probe the intelligence of our animal companions, hoping to start a conversation. We are as lonely as Defoe's Robinson Crusoe. We desperately want someone else to talk to.

The search for extraterrestrial intelligence—SETI—began fifty years ago with a lone radio astronomer borrowing spare telescope time to examine a few frequencies in the direction of two nearby stars. The search today is being conducted on a continuous basis with supercomputers and sophisticated receivers like the SETI Institute's Allen Telescope Array. Today's systems search more radio space in a few minutes than was probed in SETI's first decade. Meanwhile, China has broken ground on a new 500-meter dish (with a receiving surface the size of thirty football fields, or ten times the size of the radio telescope at Arecibo), whose mission explicitly encompasses the search for other civilizations.

Astronomers are looking as well as listening. More than three hundred extrasolar planets have been detected, all but twelve in the last decade, and more than a hundred in the last two years alone. More significant, the minimum size of detectable extrasolar planets has plummeted, making it possible to identify planets with masses similar to that of Earth. Planetary discovery has gone exponential with the recent launch of NASA's Kepler spacecraft, which will examine more than a hundred thousand stars for the presence of terrestrial-sized planets. The holy grail of planet hunters isn't Jupiter-sized giants but other Earths suitable for intelligent life recognizable to us.

The search so far has been met only by a great silence, but as astronomers continue their hunt for intelligent neighbors, computer scientists are determined to create them. Artificial intelligence research has been under way for decades, and a few AIs have arguably already passed the Turing test. Apply the exponential logic of Moore's Law and the arrival of strong AI in the next few decades seems inevitable. We will have robots smart enough to talk to and so emotionally appealing that people will demand the right to marry them.

One way or another, humanity will find someone or something to talk to. The only uncertainty is where the conversations will lead. Distant alien civilizations will make for difficult exchange because of the time lag, but the mere fact of their existence will change our self-perception as profoundly as Copernicus did five centuries ago. And despite the distance, we will of course try to talk to them. A third of us will want to conquer them, a third of us will seek to convert them, and the rest of us will try to sell them something.

Artificial companions will make for more intimate conversations, not just because of their proximity but because they will speak our language from the first moment of their stirring sentience. However, I fear what might happen as they evolve exponentially. Will they become so smart that they no longer *want* to talk to us? Will they develop an agenda of their own that makes utterly no sense from a human perspective? A world shared with superintelligent robots is a hard thing to imagine. If we are lucky, our new mind children will treat us as pets. If we are very unlucky, they will treat us as food.

AI AND INTELLECTUAL MASTERY

———◦———

JOHN TOOBY AND LEDA COSMIDES

JOHN TOOBY and LEDA COSMIDES are codirectors of
the Center for Evolutionary Psychology at the University of
California, Santa Barbara, and coeditors, with Jerome H.
Barkow, of *The Adapted Mind: Evolutionary Psychology and
the Generation of Culture*.

Currently, the most keenly awaited technological development is an all-purpose artificial intelligence—perhaps even an intelligence that would revise itself and grow at an ever-accelerating rate until it enacts millennial transformations. Since the invention of artificial minds seventy years ago, computer scientists have felt on the verge of building a generally intelligent machine. Yet somehow this goal, like the horizon, keeps retreating as fast as it is approached. We think that an all-purpose artificial intelligence will—for the foreseeable future—remain elusive. But understanding why this is so will unlock other revolutions.

AI's wrong turn was assuming that the best methods for reasoning and thinking—for true intelligence—are those that can be applied successfully to any content. Equip a computer with these general methods, input some facts to apply them to, increase hardware speed, and a dazzlingly high intelligence seems fated to emerge. Yet it never materializes, and achieved levels of general AI remain too low to meaningfully compare with human intelligence.

But powerful natural intelligences do exist. How do native intelligences—like those found in humans—operate? With few exceptions, they operate by being specialized. They break off small but biologically important fragments of the universe (predator-prey interactions, color, social exchange, physical causality, alliances, genetic kinship, and so on) and engineer different problem-solving methods for each. Evolution tailors computational hacks that work brilliantly, by exploiting relationships that exist only in their particular fragment of the universe. (The geometry of parallax gives vision a depth cue; an infant nursed by your mother is your genetic sibling; two solid objects cannot occupy the same space.) These native intelligences are dramatically smarter than general reasoning, because natural selection equipped them with radical shortcuts that bypass the endless possibilities that general intelligences get lost among. Our mental programs can be fiendishly well engineered to solve some problems because they are not limited to using only those strategies that can be applied to all problems.

Lessons from evolutionary psychology indicate that developing specialized intelligences—artificial idiot savants—and networking them would eventually achieve a mosaic AI, just as evolution gradually built natural intelligences. The essential activity is discovering sets of principles that solve a particular family of problems. Indeed, successful scientific theories are examples of specialized intelligences, whether implemented culturally among communities of researchers or implemented computationally in computer models. Similarly, adding duplicates of the specialized programs that we discover in the human mind to the emerging AI network would constitute a tremendous leap toward AI. Essentially, for this aggregating intelligence to communicate with humans—for it to understand what we mean by a question or want by a request—it will have to be equipped with accurate models of the native intelligences that inhabit human minds.

Which brings us to another impending transformation: rapid and sustained progress in understanding natural minds.

For decades, evolutionary psychologists have been devoted to perpetrating the great reductionist crime: working to create a scientific discipline that progressively maps the evolved universal human mind/brain, the computational counterpart to the human genome. The goal of evolutionary psychology has been to create high-resolution maps of the circuit logic of each of the evolved programs that together make up human nature (anger, incest avoidance, political identification, understanding physical causality, guilt, intergroup rivalry, coalitional aggression, status, sexual attraction, magnitude representation, predator-prey psychology, and so on). Each of these is an intelligence specialized to solve its class of ancestral problems. The long-term ambition is to develop a model of human nature as precise as if we had the engineering specifications for the control systems of a robot. Of course, both theory and evidence indicate that the programming of the human is endlessly richer and subtler than that of any foreseeable robot.

Still, how might a circuit map of human nature radically change the situation our species finds itself in?

Humanity will continue to be blind slaves to the programs that evolution has built into our brains until we drag them into the light. Ordinarily, we inhabit only the versions of reality they spontaneously construct for us—the surfaces of things. Because we are unaware we are in a theater, with our roles and our lines largely written for us by our mental programs, we are credulously swept up in these plays (such as the genocidal drama of us versus them). Endless chain reactions among these programs leave us the victims of history—embedded in war and oppression, enveloped in mass delusions and cultural epidemics, mired in endless negative-sum conflict.

If we understood these programs and the coordinated hallucinations they orchestrate in our minds, our species could awaken from the roles these programs assign to us. Yet this cannot happen if this knowledge—like quantum mechanics—remains forever locked up in the minds of a few specialists, walled off by the years of study required to master it.

Which brings us to another interlinked transformation, which could solve this problem.

If a concerted effort is made, we could develop methods for transferring bodies of understanding—intellectual mastery—far more rapidly, cheaply, and efficiently than we do now. Universities still use medieval techniques (lecturing) in order to noisily, haphazardly, and partially transfer fragments of twenty-first-century disciplines, taking many years and spending hundreds of thousands of dollars per transfer per person. But what if people could spend four months with a specialized AI—an immersive, interactive, all-absorbing, and video game–like experience, and emerge with a comprehensive understanding of physics, or materials science, or evolutionary psychology? To achieve this, we would have to integrate technological, scientific, and entertainment innovations in several dozen areas: Hollywood post-production techniques, the compulsively attention-capturing properties of game design, nutritional cognitive enhancement, a growing map of our evolved programs (and their organs of understanding), an evolutionary psychological approach to entertainment, neuroscience-midwived brain-computer interfaces, rich virtual environments, and 3-D imaging technologies. Eventually, conceptual education will become intense, compelling, searingly memorable, and ten times faster.

A Gutenberg revolution in disseminating conceptual mastery would change everything, and—not least—would allow our species to achieve widespread scientific self-understanding. We could awaken from ancient nightmares.

AVOIDING DOOMSDAY

—◁◦▷—

ALEXANDER VILENKIN

ALEXANDER VILENKIN is a physicist, the director of the Tufts Institute of Cosmology, L. and J. Bernstein Professor in Evolutionary Science at Tufts University, and the author of *Many Worlds in One: The Search for Other Universes.*

The long-term prospects of our civilization here on Earth are very uncertain. We can be destroyed by an asteroid impact or a nearby supernova explosion, or we can self-destruct in a nuclear or bacteriological war. It is a matter not of *if* but of *when* the disaster will strike, and the only sure way for humans to survive in the long run is to spread beyond Earth and colonize the galaxy. But our chances of doing that before we are wiped out by some sort of catastrophe appear bleak.

THE DOOMSDAY ARGUMENT

The probability for a civilization to survive the existential challenges and colonize its galaxy may be small, but it is not zero, and in a vast universe such civilizations should certainly exist. We shall call them large civilizations. There will also be small civilizations, which die out before they spread much beyond their native planets.

For the sake of argument, let us assume that small civilizations do not grow much larger than ours and die soon after they reach their maximum size. The total number of individuals who lived in such a civilization throughout its entire history is then comparable to the number of people who ever lived on Earth, which is about four hundred billion people, or sixty times the planet's current population.

A large civilization contains a much greater number of individuals. A galaxy like ours has about two hundred billion stars. We don't know what fraction of stars have planets suitable for colonization, but with what seems to be a conservative estimate of 0.005 percent, we would still have about ten million habitable planets per galaxy. Assuming that each planet will reach a population similar to that of Earth, we get four million trillion individuals. (For definiteness, we focus on humanlike civilizations, disregarding the planets inhabited by little green creatures, with one thousand per square inch.) The numbers can be much higher if the civilization spreads well beyond its galaxy. The crucial question is: What is the probability p for a civilization to become large?

It takes ten million (or more) small civilizations to provide the same number of individuals as a single large civilization. Thus, unless p is extremely small (less than one in ten million), individuals live predominantly in large civilizations. That's where we should expect to find ourselves if we are typical inhabitants of the universe. Furthermore, a typical member of a large civilization should expect to live at a time when the civilization is close to its maximum size, since that is when most of its inhabitants are living. These expectations are in a glaring conflict with what we actually observe: We live either in a small civilization, or at the very beginning of a large civilization. With the assumption that p is not very small, both of these options are unlikely, which indicates that the assumption is probably wrong. If indeed we are typical observers in the universe, then we have to conclude that the probability p for a civilization to survive long enough to become large must be very tiny. In our example, it cannot be much more than one in ten million.

This is the notorious "Doomsday argument." First suggested

by the theoretical physicist Brandon Carter about thirty-five years ago, it inspired much heated debate and has often been misinterpreted.

BEATING THE ODDS

The Doomsday argument is statistical in nature. It does not predict anything about our civilization in particular. All it says is that the odds for any given civilization to grow large are very low. At the same time, some rare civilizations do beat the odds.

What distinguishes these exceptional civilizations? Apart from pure luck, civilizations that dedicate a substantial part of their resources to space colonization, start the colonization process early, and do not stop stand a better chance of long-term survival. With many other diverse and pressing needs, this strategy may be difficult to implement, and that may be one of the reasons why large civilizations are so rare. And then, there is no guarantee. Only when the colonization is well under way, and the number of colonies grows faster than they are dying out, can one declare a victory. But if we ever reach this stage in the colonization of our own galaxy, it would truly be a turning point in the history of our civilization.

WHERE ARE THEY?

One question that needs to be addressed is: Why is our galaxy not yet colonized? There are stars in the galaxy that are billions of years older than our sun, and it should take much less than a billion years to colonize the entire galaxy. So we are faced with Enrico Fermi's famous question: "Where are they?" The most probable answer, in my view, is that we may well be the only intelligent civilization not just in the galaxy but in the entire observable universe. Our cosmic horizon is set by the distance that light has traveled since the Big Bang. It sets the absolute limit to space colonization, since no civilization can spread faster than the speed of light. There are a large number of habitable planets within our horizon, but are these planets actually inhabited? Evolution of life and intelligence requires some extremely improbable events. Theoretical estimates (admittedly rather speculative) suggest that their probability is so low that the nearest planet with intelligent life may be far beyond the horizon. If this is really so, then we are responsible for a huge chunk of real estate, ninety billion light-years in diameter. Our crossing the threshold to become a space-colonizing civilization would then really change everything. It will mean the difference between a "flicker" civilization that blinks in and out of existence and a civilization that spreads through much of the observable universe and possibly transforms it.

ESCAPING THE GRAVITY WELL

———◇———

DAVID DALRYMPLE

DAVID DALRYMPLE is a researcher in MIT's Media Lab and a student at its Center for Bits and Atoms.

Since I've lived for only seventeen years, to ask what I expect to live to see is to cast a long, wide net.

When I look far into the future, I find it a useful exercise to imagine myself as a nonhuman scientist: an alien, a god, or some other creature with a modern understanding of mathematics and physics but no inherent understanding of human culture or language beyond what I can deduce from watching what happens from on high. Essentially, such a creature would look at the world "up to isomorphism": It is not relevant who does what, what the observed call themselves, whether they have ten fingers or twelve, but rather how many of these creatures there are, whether they survive, and where they go.

From that perspective, a few things are apparent: We are depleting our planet's resources faster than they can be replenished. Most of the sun's energy is reflected back into space without being used. There are more of us every minute, and we don't seem to be effectively slowing our growth, despite overcrowding and lack of resources. We are trapped in a delicate balance of environmental conditions that has been faltering ever since we began pulling hydrocarbons out of the crust and burning them in the atmosphere, and there seems to be a good chance that

that balance will tip catastrophically in the next hundred years, if we don't run out of hydrocarbons first. We have thrown countless small, special-purpose objects into space, and some have transmitted very valuable information back to us. For a short period (while it seemed that we would destroy our planet with deliberate nuclear explosions and immediate evacuation might be necessary), we played at shooting living men into the sky, but they have gone only as far as our planet's moon, still within Earth orbit, and, sure enough, wound up right back in our atmosphere. I should note here that I do not mean any disrespect for the achievements of the Apollo program; in fact, I believe they are among humankind's greatest—so far!

If civilization is to continue expanding, however—as well it shall if it does not collapse—it must escape the tiny gravity well it is trapped in. It is quite unclear to me how this will happen: whether humans will look anything like the humans of today, whether we will escape to sun-orbiting space stations or planetary colonies, but if we expand, we must expand beyond Earth. Even if environmentalists succeed in building a sustainable terrestrial culture around local farming and solar energy, it will remain sustainable only if we limit reproduction, which I expect most of modern society to find unconscionable on some level.

It has always been not only the human way but the way of all living things to multiply and colonize new frontiers. What is uniquely human is our potential ability to colonize *all* frontiers—to adapt our intelligence to new environments or adapt environments to suit ourselves. Although the chaos of a planetary atmosphere filled with organic diversity is a beautiful and effective cradle of life, it is no place for the new human/machine civilization. By some means—genetic engineering, medical technology, brain scanning, or something even more fantastical—I expect that humans will gradually shorten the food chain, adapting to use more directly the energy of stars. Perhaps we will genetically modify humans to photosynthesize directly, or implant devices that can provide all the energy for the necessary chemical reactions electrically, or scan our intelligences into solar-powered computing devices. Again, the details are hard to predict, but I believe there will be some way forward.

I'm getting ahead of myself, so let's come back to the present. There is budding new interest in the development of space technology—in large part undertaken as private ventures, unlike in the past. Many view such operations as absurd luxury vacations for the superrich, or at best as unlikely schemes to harvest fuel on the moon and ship it back to Earth via a rail gun. I believe this research is tremendously important, because whatever short-term excuses may be found to fund it, in the long run it is absolutely critical to the future development of our civilization. I also don't mean to imply that we should give up on environmentalism and sustainability and just start over with another planet; these principles will only become more important as we spread far and wide, beginning in each new place with even more limited resources and limited contact with home. Not to mention that if Earth can be saved, it would be a tremendous cultural treasure to preserve as long as possible.

I'm not as optimistic about interstellar travel as some (I certainly don't expect it to become practical in this century), but I'm also much more optimistic about the ability of human civilization to adapt and survive without the precise conditions that were necessary for its evolution. There are so many possible solutions for the survival of humans (or posthumans) in solar orbit or on "inhospitable" planets that I expect we will find some way to make it work long before generational or faster-than-light voyages to faraway star systems. In fact, I expect it in my lifetime. But someday "escaping the gravity well" will mean not that of Earth but that of our star, and then humankind's ship will at last have . . . gone out.

SYNTHETIC BIOLOGY
WITH INTERPLANETARY REACH

———◇———

DIMITAR SASSELOV

DIMITAR SASSELOV is an astrophysicist at Harvard University.

To an astrophysicist, "everything" is the universe. What could we possibly do to change that "everything"? People like to say that a scientific idea changed the world when Copernicus suggested that the sun, not Earth, is at the center. People regard the invention of the Internet as a development that changed the world. The list is long.

But which "world" did these ideas change? Well, yes, they are all about us, *Homo sapiens*, a recently evolved branch on the tree of life, with roots in a biochemistry that somehow, four billion years ago, took hold on planet Earth. And yes, *Homo sapiens* has created amazing things: airplanes, antibiotics, phones, computers, but none of these will change the orbits of the stars.

And so it was, until now. There is a game-changing scientific development that transcends all in human history. It is already under way, and it even has a name: synthetic biology. People take synthetic biology to mean different things. Most often, it's reduced to synthetic genomics, redesigning the genomes of organisms to make them act in new ways—such as microbes that produce fuel or pharmaceutical products. I take synthetic biology to mean creating new trees of life, as opposed to engineering new branches to the existing tree. In my view, synthetic biology is about engineering an alternative biochemistry, "seeding" an alternative tree that then evolves on its own. In that alternative, the life is as natural as any life we know. I shall let others describe it and explain how they will do it. One thing is sure: Biologists will use synthesis the same way chemists do today—routinely.

But there is more! It is the interplanetary reach of synthetic biology that will make it a new phenomenon in the cosmos we know. Life is a planetary phenomenon that can transform a planet. The development of synthetic biology appears to be a stage in life's evolution whereby some of its forms can leave the host planet and adapt to other environments, potentially transforming other planets and eventually the galaxy.

LIFE (OR NOT) ON MARS

---◄○►---

RODNEY BROOKS

RODNEY BROOKS is Panasonic Professor of Robotics at MIT, cofounder of iRobot Corp., chief technical officer and chairman of Heartland Robotic, and the author of *Flesh and Machines: How Robots Will Change Us.*

I am very sure that in my lifetime we will have a definitive answer to one question that has been debated, with little data, for hundreds of years. The answer as to whether or not there is life on Mars will either be a null result, if negative, or it will profoundly affect science (and perhaps philosophy and religion) if positive.

As 1990s NASA administrator Dan Goldin rightly reasoned, the biggest possible positive public-relations coup for his agency (and therefore for its continued budget) would be its discovery of unambiguous evidence of life elsewhere in the universe. One of the legacies of that judgment is the almost weekly discovery of new planets orbiting nearby stars. If life does exist outside our solar system, the easy bet is that it exists on planets, so we had better find planets to look at for direct evidence of life. We have been able to infer the existence of very large planets by carefully measuring star wobbles; more recently, we have detected smaller planets by measuring their occultations—the way they dim their star as they pass between it and Earth. And just in the last months of 2008, we had our first direct images of planets orbiting other stars.

NASA has an ambitious program—using the Hubble and Spitzer Space Telescopes and the Terrestrial Planet Finder (to be launched in 2016)—to get higher-and-higher-resolution images of extrasolar planets and look for telltale chemical signatures of large-scale biochemical effects of Earthlike life on them. If these attempts do indeed discover life elsewhere, it will greatly alter our views of life and no doubt stimulate much creative thinking that will lead to new science about Earth's version of life. But it will take us a long, long time to infer the nature of that distant life in significant detail and the similarities and differences to our own.

The second of Goldin's legacies concerns life much closer to home. NASA has a strong—but somewhat endangered at this moment—direct exploration program for the surface of Mars. We have not yet found direct evidence of life there, but neither have the options for its existence narrowed appreciably. And we are very rapidly learning much more about likely locations for life; again, just in the last months of 2008 we discovered vast water glaciers with only a shallow covering of soil. We have many more exciting places to explore for life on Mars than we will be able to send probes to over the next few years. If we do discover Martian life (alive or extinct), one can be sure there will be a flurry of missions launched to examine it or its remnants in great detail.

There are a range of possibilities for what Martian life might look like—and consequent ambiguity about whether its creation was independent of that on Earth or there was cross-contamination of these two planets and only one genesis for life.

At one extreme, life on Mars could turn out to be DNA-based with exactly the same coding scheme for amino acids that all life on Earth uses. Or it could look like a precursor to Earth life, sharing a compatible precursor encoding—perhaps an

RNA-based life-form or even a PNA (peptide nucleic acid)-based form. Any of these outcomes would help us immensely in our understanding of the development of life from nonlife, whether it happened on Mars or Earth.

Another possible discovery would be one of these forms with a different, or even an incompatible, encoding for known amino acids. That would be a far more radical outcome. It would tell us two things: Life arose twice, spontaneously and separately, on two adjacent planets in one particular little solar system. The universe must, in that case, be teeming with life. But more than that, such a discovery would tell us that the space of possible biochemistries is probably rather narrow—and that, in turn, immediately tells us a lot about all those other life-forms out there. We will also have an idea about the probable spaces we should be searching in our efforts to synthesize new life-forms in the laboratory.

The most mind-expanding outcome would be if the life we discover on Mars is not at all based on a genetic coding scheme of long chains of purine bases that decode in triplets to select an amino acid to be tacked onto a protein under construction. This would revolutionize our understanding of the possibilities for biology. It would provide us with a completely different form to study. It would expand the possibilities for what must be invariant in biology and what can be manipulated and engineered. It would completely change our understanding of ourselves and our universe.

A SEPARATE ORIGIN FOR LIFE

———◈———

ROBERT SHAPIRO

ROBERT SHAPIRO is professor emeritus of chemistry and senior research scientist at New York University and the author of *Planetary Dreams: The Quest to Discover Life Beyond Earth*.

We may find the evidence we need within the frigid hydrocarbon lakes of Titan. Or perhaps we shall locate it by tracing the source of the methane hot spots on Mars to their deep underground origin. It may be easier, though, to sample in depth the contents of the water vapor and ice jets that erupt from "tiger stripe" cracks near the south pole of Enceladus. Least expensive of all would be to explore closer to home—in forsaken regions of Earth that are so hot, or so acidic, or so lacking in some vital nutrient necessary to life as we know it, that no creatures built of such life would deign to inhabit it. So many promising leads have appeared that it seems likely that only our indifference and our finances stand as obstacles to our gaining the prize.

The prize in this case would be a sample of truly alien life. Despite the great legacy from pulp science-fiction magazines and expensive Hollywood productions, aliens need not be green men or menacing fanged monsters. Even humble microbes dismissed as "shower scum" in *The New York Times* would do nicely, provided they met one key requirement: They must differ enough from us at the biochemical level, so that it would be clear that they had started up and evolved on their own. Two separate origins in the same solar system would imply that the universe is liberally sprinkled with life.

Why would a discovery of this type change everything? It would not put food on our dinner tables or shorten our commute to work. The change would come in our perception of the universe and the place of life within it. We would learn that life, like art, can take on many forms and be constructed in countless ways, and that we appear to be residents of a universe built to encourage such diversity.

We have always understood, of course, that living things came in many sizes and shapes: bacteria and whales, octopuses and centipedes. But we took it for granted that they were all made of one substance, one flesh. When they were ferocious, they could devour us. When they were domestic, we could make meals of them. This expectation was extended to alien life in fiction, myth, and imagination. The Martian invaders of H. G. Wells's *War of the Worlds* were ultimately subdued by infection by Earthly microbes. The creatures of the *Aliens* film series could incubate in humans and draw nourishment from them. Humans could have sexual encounters with intruders in flying saucers just as they did with the gods of Ancient Greece.

Such events would have provoked little surprise in the nineteenth century, when the basic substance of life was thought to be a vital gel-like protoplasm, which presumably would be the same everywhere. We now understand that life's basic structure is much more intricate, but that the same building materials— nucleic acids, proteins, carbohydrates, and fatty substances—are used by all known life-forms here. Some scientists have extended this conclusion to alien life. To quote Nobel laureate George Wald, "So I tell my students: Learn your biochemistry here and you will be able to pass examinations on Arcturus." This view carries practical consequences today. Proposals for instruments to be flown to Mars and elsewhere include antibodies, probes, and other methods designed to detect the molecules familiar to us on Earth today. If the microbes we discovered on a nearby world had traveled from Earth inside a meteorite, then this expectation would be valid. By studying them, we might learn a lot about the earlier stages of the evolution of life here, and about the ability of Earth life to adapt to a very different environment. Such information would be valuable, but it would not change everything.

When we discover a separate origin for life, we will have hit the jackpot. It will truly be made of a different flesh. Biochemists will be fascinated to learn how life functions such as energy capture, information storage, and catalysis can be carried out by materials different from the life we know. The field of biology will be greatly enriched, and a host of new technical innovations may arise from the new knowledge—but even this would not change everything. The largest change would be in the way we view our existence and plan for our future.

For stability and comfort, most human beings appear to require a narrative that provides meaning and purpose to their lives. In many religions, one's behavior determines one's fate in a hereafter. One's actions are crucial in the grand scheme of things. Prior to the Copernican revolution, Earth was naturally placed center stage in the theater of existence. As suitable decorations, the various heavenly bodies were embedded in spheres that rotated above us. Now we understand that our home world occupies only a minute fragment of an immense array of planets, stars, and galaxies. Our species has experienced only a sliver of the great expanse of time that has passed since the Big Bang, and much more is yet to come. The playing field has become immense.

Traditional religions have generally ignored this huge expansion of the cosmic scheme and cling to an essentially pre-Copernican view of existence. In doing so, they reduce their narratives to cherished folk tales, with a message as relevant today as the science of Aristotle. By contrast, some Nobel laureate scientists have regarded the universe as meaningless and pointless, with our life representing an accidental anomaly that will disappear sooner or later.

Fortunately, another interpretation exists—one that is fully compatible with science though it extends beyond it. Eric Chaisson, Paul Davies, and others have described a viewpoint often called cosmic evolution. The successive appearance of galaxies, stars, planets, atoms and molecules, life and intelligence is seen as inherent in the laws that have governed our universe since the Big Bang. Minor alterations in many of the fundamental constants embedded in those laws would have made this succession of events impossible. For whatever reason, our universe is (to use Paul Davies' word) biofriendly.

If a separate origin of life were encountered within our own solar system, the credibility of this viewpoint would be strengthened immensely. We could see ourselves as active participants in a vast cosmic competition—one that required all available space and time to play itself out to the fullest extent. To advance in the game and ultimately grasp its point, our mission would be to survive, prosper, evolve to the next stage, and expand into the greater universe—not bad goals under any circumstances. By liberating humanity from a choice between obsolete dogma and unrelenting pessimism, this discovery would ultimately change everything.

SHADOW BIOSPHERE

◄○►

PAUL DAVIES

PAUL DAVIES is a physicist at Arizona State University, director of Beyond: The Center for Fundamental Concepts in Science, and the author of *Cosmic Jackpot: Why Our Universe Is Just Right for Life*.

A hundred and fifty years ago, Charles Darwin gave us a convincing theory of how life has evolved, over billions of years, from primitive microbes to the richness and diversity of the biosphere we see today. But he pointedly left out of account how life got started in the first place. "One might as well speculate about the origin of matter," he quipped in a letter to a friend. How, where, and when life began remain some of the greatest unsolved problems of science. Even if we make life in the laboratory in the near future, it still won't tell us how Mother Nature did it without expensive equipment, trained biochemists and—the crucial point—a preconception of the goal to be achieved. However, we might be able to discover the answer to a more general question: Did life originate once, or often?

The subject of astrobiology is predicated on the hope and expectation that life emerges readily in Earthlike conditions and is therefore likely to be widespread in the universe. The assumption that, given half a chance, life will out, is sometimes called biological determinism. Unfortunately, nothing in the known laws of physics and chemistry singles out the state of matter we call "living" as in any way favored. There is no known law that fast-tracks matter to life. If we do find life on another planet and we can be sure it has started there from scratch, completely independent of life on Earth, biological determinism will be vindicated. With NASA scaling back its activities, however, the search for extraterrestrial life has all but stalled.

Meanwhile, there is an easy way to test biological determinism right here and now. No planet is more Earthlike than Earth itself, so biological determinism predicts that life should have started many times on our home planet. That raises the fascinating question of whether there might be more than one form of life inhabiting the terrestrial biosphere. Biologists are convinced that all known species belong to the same tree of life and share a common origin. But almost all life on Earth is microbial, and only a tiny fraction of microbes have been characterized, let alone sequenced and positioned on the universal tree. You can't tell by looking what makes a microbe tick; you have to study its innards. Microbiologists do that using techniques carefully customized to life as we know it. Their methods wouldn't work for an alternative form of life. If you go looking for known life, you are unlikely to find unknown life.

I believe there is a strong likelihood that Earth possesses a shadow biosphere of alternative microbial life representing the evolutionary products of a second genesis. Maybe also a third, fourth . . . . I also think we might very well discover this shadow biosphere soon. It could be ecologically separate, confined to niches beyond the reach of known life by virtue of extreme heat, cold, acidity, or other variables. Or it could interpenetrate the known biosphere in both physical and parameter space. There could be, in effect, alien microbes right under our noses (or even in our noses). Chances are, we would not be aware of the fact, especially if the weird shadow life is present at relatively low abundance. But a targeted search for weird microbes and the weird viruses that prey on them could find shadow life any day now.

Why would that change everything? Apart from the sweeping technological applications that having a second form of life would bring, the discovery of a shadow biosphere would prove biological determinism and confirm that life is indeed widespread in the universe. To expect that life would start twice on Earth but never on another planet like Earth is too improbable. And if the universe is teeming with life, it is far more likely that there is also intelligent life elsewhere in the universe. We might then have greater confidence that the answer to the biggest of the big questions of existence—"Are we alone in the universe?"—is very probably No.

LABORATORY EARTH COLONIES

---◄◊►---

JOHN GOTTMAN

JOHN GOTTMAN is a psychologist, founder of the Gottman Institute, and the coauthor, with Julie Gottman, of
*And Baby Makes Three: The Six-Step Plan for Preserving
Marital Intimacy and Rekindling Romance After Baby Arrives.*

The technological changes were small at first. In 2007,
a telescope was developed that could search for planets in the
Milky Way within one hundred light-years of Earth. The next
version of the telescope, in 2008, did not have to block out the
light of the new star to see the planets; it could directly see the
reflected light of the planets closest to every star. That made it
possible to do spectroscopic analysis of reflected light and search
for blue planets like Earth. Within a decade, a hundred Earthlike planets had been identified within a hundred light-years. In
the next two centuries, that number increased to fifty thousand
blue planets.

Within the next two centuries, the seemingly impossible
technical problems of space travel began to be solved. The
moon, Europa, and Mars were colonized. Terraforming technologies were developed. Problems of foil sails were solved. Designs emerged within two years for ships that could get up to 85
percent of the speed of light, using acceleration from starlight
and harnessing the energy of empty space itself. Many designs

emerged for a spinning 2-mile-diameter Earth habitat ship that produced a 1-g environment. Thousands of people wanted to make trips around the galaxy.

Laboratory Earth colonies were formed for simulating conditions on the galactic trips. From these experiments, social scientists soon realized that the major unsolved problem of galactic colonization was the social-psychological problem: How could humans live together for up to fifty-two years, raising children who would become the explorers of the blue planets? Much had been learned, of course, from the social-psychological studies in the twenty-first and twenty-second centuries that aimed at obtaining planetwide cooperation in combating global warming, establishing sustainable energy production, and ending worldwide hunger and disease. But that work was primitive and rudimentary compared with the challenges of galactic colonization.

The subsequent classic social-psychological studies were all funded privately by one man. Thousands of scientists participated. Studies of all kinds were initially devised, and the results were carefully replicated. The entire series of social-psychological experiments took a century to perform. It rapidly became clear that a military—or any hierarchical—social structure could not last without external threat; the work of anthropologist Peggy Sanday back in the late-twentieth and early twenty-first centuries had demonstrated that fact without question. The problem was how to foster creative collaboration and minimize self-interest. Eventually, it was deemed necessary for each ship to spend five years prior to the trip selecting a problem that all the members would creatively and cooperatively face. The work would have to easily consume the ship's crew for sixty years. In addition, each ship would represent a microcosm of all Earth's activities, including all the occupations and professions, adventure, play, and sports.

In the year 2500, more than twenty thousand ships set out, two headed for each planet. Inevitably, many ships would successfully make the journey. No one knew what they would find. There was no plan for communication between the stars. The colonization of the Milky Way had begun.

INTERSTELLAR VIRUSES

—◇—

GEORGE DYSON

GEORGE DYSON is a science historian and the author of *Darwin Among the Machines: the Evolution of Global Intelligence.*

The detection of extraterrestrial life, extraterrestrial intelligence, or extraterrestrial technology (there's a difference) will change everything. The game could be changed completely by an extraterrestrial presence discovered (or perhaps not discovered) here on Earth.

SETI@home, our massively distributed search for extraterrestrial communication, now links some five million terrestrial computers to a growing array of radio telescopes, delivering a collective 500 teraflops of fast Fourier transforms representing a cumulative two million years of individual processing time. Not a word (or even a picture) so far.

However, as AI pioneer Marvin Minsky warned in 1970: "Instead of sending a picture of a cat, there is one area in which you can send the cat itself." Life, assuming it exists elsewhere in the universe, will have had time to explore an unfathomable diversity of forms. Those best able to survive the passage of time, adapt to changing environments, and migrate unscathed across interstellar distances will become the most widespread. Life-forms that assume digital representation for all or part of their life cycle will not only be able to send *messages* at the speed of light, they will be able to send *themselves*.

Digital organisms can be propagated economically even with extremely low probability of finding a host environment in which to germinate and grow. If the kernel is intercepted by a host that has discovered digital computing—whose ability to translate between sequence and structure (as Alan Turing and John von Neumann demonstrated) is as close to a universal common denominator as life and intelligence running on different platforms may be able to get—it has a chance. If we discovered such a kernel, we would immediately replicate it widely. Laboratories all over the planet would begin attempting to decode it, eventually compiling the coded sequence—intentionally or inadvertently— to utilize our local resources, the way a virus is allocated privileges within a host cell. The read-write privileges granted to digital organisms already include material technology, human minds, and, increasingly, biology itself. (What, exactly, *are* those screen savers doing at Dr. Venter's laboratory during the night?)

According to Edward Teller, Enrico Fermi once asked "Where is everybody?" when the subject of extraterrestrial beings came up over lunch at Los Alamos in 1950. The answer may be "We've arrived! Now help us unpack!" Fifty years later, over lunch at Stanford, I asked a ninety-one-year-old Edward Teller (holding a wooden staff at his side like an Old Testament prophet) how Fermi's question was holding up.

"Let me ask you," Teller interjected in his thick Hungarian accent. "Are you uninterested in extraterrestrial intelligence? Obviously not. If you are interested, what would you look for?"

"There's all sorts of things you can look for," I answered. "But I think the thing not to look for is some intelligible signal. Any civilization that is doing useful communication, any efficient transmission of information, will be encoded, so it won't be intelligible to us. It will look like noise."

"Where would you look for that?" asked Teller.

"I don't know."

"I do!"

"Where?"

"Globular clusters!" answered Teller. "We cannot get in touch with anybody else, because they choose to be so far away from us. In globular clusters, it is much easier for people at different places to get together. And if there is interstellar communication at all, it must be in the globular clusters."

"That seems reasonable," I agreed. "My own personal theory is that extraterrestrial life could be here already—and how would we necessarily know? If there is life in the universe, the form of life that will prove to be most successful at propagating itself will be digital life. It will adopt a form that is independent of the local chemistry and migrate from one place to another as an electromagnetic signal, as long as there's a digital world—a civilization that has discovered the Universal Turing Machine—for it to colonize when it gets there. And that's why von Neumann and you other Martians got us to build all these computers—to create a home for this kind of life."

There was a long, drawn-out pause. "Look," Teller finally said, lowering his voice, "may I suggest that instead of explaining this, which would be hard, you write a science-fiction book about it?"

"Probably someone has," I said.

"Probably," answered Teller, "someone has not."

COMPUTERS ARE THE NEW MICROSCOPES

---◁◦▷---

TERRENCE SEJNOWSKI

TERRENCE SEJNOWSKI is a computational neuroscientist at the Salk Institute for Biological Studies and the co-author, with Patricia Churchland, of *The Computational Brain*.

Scientific ideas change when new instruments are developed that detect something new about nature. Electron microscopes, radio telescopes, and patch recordings from single-ion channels have all led to game-changing discoveries.

We are in the midst of a technological revolution in computing that has been unfolding since 1950 and is having a profound impact on all areas of science and technology. As computing power doubles every eighteen months according to Moore's Law, unprecedented levels of data collection, storage, and analysis have revolutionized many areas of science.

For example, optical microscopy is undergoing a renaissance, as computers have enabled us to localize single molecules with nanometer precision and image the extraordinarily complex molecular organization inside cells. This has become possible because computers allow beams to be formed and photons to be collected over long stretches of time, perfectly preserved and processed into synthetic pictures. High-resolution movies are revealing the dynamics of macromolecular structures and molecular interactions for the first time.

In trying to understand brain function, we have until recently relied on microelectrode technology that limited us to recording from one neuron at a time. Coupled with advances in molecular labels and reporters, new 2-photon microscopes, guided by computers, will soon make it possible to image the electrical activity and chemical reactions occurring inside millions of neurons simultaneously. This will realize Charles S. Sherrington's dream of seeing brain activity as an "enchanted loom where millions of flashing shuttles weave a dissolving pattern, always a meaningful pattern though never an abiding one; a shifting harmony of subpatterns."

Computers have become the new microscopes, allowing us to see behind the curtains. By 2015, their power will begin to approach the neural computation that occurs in brains. This does not mean we will be able to understand that computation, only that we can begin to approach the complexity of a brain on its own terms. Coupled with advances in large-scale recordings from neurons, this increase in computer power should by then enable us to crack many of the brain's mysteries, such as how we learn and where memories reside—though I would not expect a computer model of human-level intelligence to emerge without other breakthroughs that cannot be predicted.

SILICON IMMORTALITY: DOWNLOADING CONSCIOUSNESS INTO COMPUTERS

DAVID EAGLEMAN

DAVID EAGLEMAN is a neuroscientist at Baylor College of Medicine and the author of *Sum: Forty Tales from the Afterlives*.

While medicine will advance in the next half century, we are not on course to achieve immortality by curing all disease. Bodies simply wear down with use. We are on course, however, to achieve technologies that let us store unthinkable amounts of data and run gargantuan simulations. Therefore, well before we understand how brains work, we will find ourselves able to digitally copy the brain's structure and download the conscious mind into a computer.

If the computational hypothesis of brain function is correct, it suggests that an exact replica of your brain will hold your memories, will think and feel the way you do, and will experience your consciousness—irrespective of whether it is built of biological cells, Tinkertoys, or zeros and ones. The important part about brains, the theory goes, is not the structure; it is the algorithms that ride on top of the structure. So if the scaffolding that supports the algorithms is replicated—even in a different medium—then the resultant mind should be identical. If this proves correct, it is almost certain that we will soon have technologies that allow us to copy and download our brains and live forever in silica. We will not have to die anymore. We will instead inhabit virtual worlds like that of *The Matrix*. I assume there will be markets for purchasing various kinds of afterlives and sharing them with other people—this is the future of social networking. And once you are downloaded, you may even be able to watch the death of your outside, real-world body, just as we might watch an interesting movie.

This hypothesized future embeds many assumptions, the speciousness of any one of which could spill the house of cards. The main problem is that we don't know exactly which variables are critical to capture in our hypothetical brain scan. Presumably the important data will include the detailed connectivity of the hundreds of billions of neurons. But knowing the point-to-point circuit diagram of the brain may not be sufficient to specify its function. The exact three-dimensional arrangement of the neurons and glia is likely to matter as well (for example, because of three-dimensional diffusion of extracellular signals). We may further need to probe and record the strength of each of the trillions of synaptic connections. In a still-more-challenging scenario, the states of individual proteins (phosphorylation states, exact spatial distribution, articulation with neighboring proteins, and so on) will need to be scanned and stored. It should also be noted that a simulation of the central nervous system by itself may not be sufficient for a good simulation of experience: Other aspects of the body may require inclusion, such as the endocrine system, which sends and receives signals from the brain. These considerations potentially lead to billions of trillions of variables that need to be stored and emulated.

The other major technical hurdle is that the simulated brain must be able to modify itself. We need not only the pieces and parts, but also the physics of their ongoing interactions—for example, the activity of transcription factors that travel to the nucleus and cause gene expression, the dynamic changes in location and strength of the synapses, and so on. Unless your simulated experiences change the structure of your simulated brain, you will be unable to form new memories and will have no sense of the passage of time. Under those circumstances, is there any point in immortality?

The good news is that computing power is blossoming sufficiently quickly that we are likely to make it within a half century. And note that a simulation does not need to be run in real time in order for the simulated brain to believe it is operating in real time.

There is no doubt that whole-brain emulation is an exceptionally challenging problem. As of this moment, we have no neuroscience technologies geared toward ultra-high-resolution scanning of the sort required—and even if we did, it would take several of the world's most powerful computers to represent a few cubic millimeters of brain tissue in real time. It's a large problem. But assuming that we haven't missed anything important in our theoretical frameworks, then we have the problem cornered, and I expect to see the downloading of consciousness come to fruition in my lifetime.

THE IMPLEMENTATION OF LIFE IN ENGINEERED MATERIALS

———◦———

NEIL GERSHENFELD

NEIL GERSHENFELD is a physicist at MIT and the author of *FAB: The Coming Revolution on Your Desktop—From Personal Computers to Personal Fabrication*.

Life is defined by organic chemistry. There's software for artificial life and artificial intelligence, but these are, well, artificial—they exist *in silico* rather than in vivo. Conversely, synthetic biology is recoding genes, but it isn't very synthetic; it uses the same sets of proteins as the rest of molecular biology. If, however, bits could carry mass as well as information, the distinction between artificial and synthetic life would disappear. Virtual and physical replication would be equivalent.

There are in fact promising laboratory systems that can compute with bits represented by mesoscopic materials rather than electrons or photons. Among the many reasons to do this, the most compelling is fabrication: Instead of a code controlling a machine to make a thing, the code can itself become a thing (or many things).

That sounds a lot like life. Indeed, current work is developing micron-scale-engineered analogs to amino acids, proteins, and genes, a "millibiology" to complement the existing microbiology. By working with components that have macroscopic physics but microscopic sizes, the primitive elements can be selected for their electronic, magnetic, optical, or mechanical properties as well as active chemical groups.

Biotechnology is booming. But it is very clearly segregated from other kinds of technology that contribute to the study of, but not the identity of, biology. If, however, life is understood as an algorithm rather than a set of amino acids, then the creation of such really-artificial or really-synthetic life can enlarge the available materials, length, and energy scales. In such a world, biotechnology, nanotechnology, information technology, and manufacturing technology merge into a kind of universal technology of embodied information. Beyond the profound practical implications, forward- rather than reverse-engineering life may be the best way to understand it.

DECODING THE BRAIN

———◦———

GARY MARCUS

GARY MARCUS is a professor of psychology at New York University and the author of *Kluge: The Haphazard Construction of the Human Mind*.

Within my lifetime (or soon thereafter), scientists will finally decode the language of the brain. At present, we understand a bit about the basic alphabet of neural function, how neurons fire, and how they come together to form synapses, but we haven't yet pieced together the words, let alone the sentences. Right now, we're sort of like Gregor Mendel at the dawn of genetics: He knew there must be something like genes (what he called "factors") but couldn't say where they lived (in the protein? in the cytoplasm?) or how they got their job done. Today we know that thought has something to do with neurons and that our memories are stored in brain matter, but we don't yet know how to decipher the neurocircuitry.

Doing that will require a quantum leap. The most popular current techniques for investigating the brain, like functional magnetic resonance imaging (fMRI), are far too coarse. A single three-dimensional "voxel" in an fMRI scan lumps together the actions of tens or even hundreds of thousands of neurons, yielding a kind of rough geography of the brain (emotion in the amygdala, decision making in the prefrontal cortex) but little in the way of specifics. How does the prefrontal cortex actually do its thing? How does the visual cortex represent the difference between a house and a car, or a Hummer and a taxi? How does Broca's area know the difference between a noun and verb? To answer questions like these, we need to move beyond the broad-scale geographies of fMRI, down to the level of individual neurons.

At the moment, that's a big job. For one thing, in the human brain there are billions of neurons and trillions of connections between them; the sheer amount of data involved is overwhelming. For another, until recently we've lacked the tools to understand the function of individual neurons in action, within the context of microcircuits.

But there's good reason to think all that is about to change. Computers continue to advance at a dizzying pace. Then there's the truly unprecedented explosion in databases such as the Human Genome Project and the Allen Brain Atlas—enormously valuable data sets that are shared publicly and are instantly available to all researchers everywhere. Even a decade ago, there was nothing like them. Finally, genetic neuroimaging is just around the corner: Scientists can now induce individual neurons to fire and (literally) light up on demand, allowing us to understand individual neural circuits in a brand-new way.

Technical advances alone won't be enough, though. We'll need a scientist with the theoretical vision of Francis Crick, who not only helped identify the physical basis of genes—the double helix of DNA—but also the code by which the individual nucleotides of a gene get translated (in groups of three) into amino acids, the building blocks of proteins. When it comes to the brain, we already know that neurons are the physical basis of thinking and knowledge, but we don't know the laws of translation that relate neurons to information.

I don't expect that there will be one single code. Although every creature uses essentially the same translation between DNA and amino acids, different parts of the brain may translate between neurons and information in different ways. Circuits that control muscles, for example, seem to work on a system of statistical averaging; the angle at which a monkey extends its arm seems, as best we can tell, to be a kind of statistical average of the actions of hundreds of individual neurons, each representing a slightly different angle of possible motion—44 degrees, 44.1 degrees, and so forth. Alas, what works for muscles probably can't work for sentences and ideas—so-called declarative knowledge, like the proposition that "Michael Bloomberg is the mayor of New York" or the idea that my flight to Montreal leaves at noon. It is implausible that the brain would have vast population of neurons reserved for each specific thought I might entertain ("My flight to Montreal leaves at 11:58 a.m.," "My flight to Montreal leaves at 11:59 a.m.," and so on). Instead, the brain, like language itself, needs some sort of combinatorial code—a way of putting pieces (Montreal, flight, noon) together into larger elements.

When we crack that nut—when we figure out how the brain manages to encode declarative knowledge—an awful lot is going to change. For one thing, our relationship to computers will be completely and irrevocably altered; clumsy input devices like mice, windows, keyboards, and even heads-up displays and speech recognizers will go the way of typewriters and fountain pens. Our connection to computers will be far more direct. Education, too, will fundamentally change, as engineers and cognitive scientists begin to leverage an understanding of brain code into ways of directly uploading information into the brain. Knowledge will become far cheaper than it already has in the Internet era. With luck and wisdom, we as a species could advance immeasurably.

CHEAP CRYONIC SUSPENSION OF BRAINS

—◦—

BART KOSKO

BART KOSKO is an information scientist at the University of Southern California and the author of *Noise*.

Society will change when the poor and middle class have easy access to cryonic suspension of their cognitive remains—even if the future technology involved ultimately fails.

Today we almost always either bury dead brains or burn them. Both disposal techniques result in irreversible loss of personhood information, because both techniques either slowly or quickly destroy all the brain tissue that houses a person's unique neural-net circuitry. The result is a neural information apocalypse, with all the denial and superstition that every culture has evolved to cope with it.

Some future biocomputing technology may extract and thus back up this defining neural information or wetware. But no such technology is in sight despite the steady advances of Moore's Law doubling of transistor density on computer chips every two years or so. Nor have we cracked the code of the random pulse train from a single neuron. Hence we are not even close to making sense of the interlocking pulse trains of the billions of chattering neurons in a functioning human brain.

So far the only practical alternative to this information catastrophe is to vitrify the brain and store it indefinitely in liquid nitrogen at about $-320°\,F$. Even the best vitrification techniques still produce massive cell damage that no current or even medium-term technology can likely reverse. But the shortcomings of early-twenty-first-century science and engineering hardly foreclose the technology options that will be available in a century, and far less so in a millennium. Suspended brain tissue needs only periodic replacement of liquid nitrogen to wait out the breakthroughs.

Yet right now there are only about a hundred brains suspended in liquid nitrogen in a world where each day about a hundred and fifty thousand people die. That comes to fewer than three suspended brains per year since a forty-year-old, post–*Space Odyssey* Stanley Kubrick hailed the promise of cryonic suspension in his 1968 *Playboy* interview. Kubrick cast death as a problem of bioengineering: "Death is no more natural or inevitable than smallpox or diphtheria. Death is a disease and as susceptible to cure as any other disease." The *Playboy* interviewer asked Kubrick if he was interested in being frozen. Kubrick said that he "would be if there were adequate facilities available." But just over three decades later, Kubrick opted in effect for the old neural apocalypse when he could easily have afforded a first-class cryonic suspension in quite adequate facilities.

The Kubrick case shows that dollar cost is just one factor that affects the ease of mass access to cryonics. Today many people can afford a brain-only suspension by paying moderate premiums for a life-insurance policy that would cover the expenses. But almost no one accepts that cryonics wager. There are also stigma costs from the usual scolds in the church and in bioethics. There is likewise no shortage of biologists who will point out that you cannot get back the cow from the hamburger.

And there remains the simple denial of the inexorable neural catastrophe. That denial is powerful enough to keep most citizens from engaging in rational estate planning. The probate code in some states such as California even allows valid handwritten wills that an adult can pen (but not type) and sign in minutes and without any witnesses. But only a minority of Californians ever executes these handwritten wills or the more formal attested wills. The great majority die intestate, and thus they let the chips fall where the state says they fall.

So it is not too surprising that the overwhelming majority of the doomed believe that the real or imagined transaction costs of brain suspension outweigh its benefits—if they think about the matter at all. But those costs will only fall, as technology marches on ever faster and as the popular culture adapts to those tech changes. One silver lining of the numbing parade of comic-book action movies is how naturally the younger viewing audience tends to embrace the fanciful information and biotechnology involved in such fare, even if they lack a like enthusiasm for calculus.

Again, none of this means that brain suspension in liquid nitrogen will ever work in the sense that it leads to some type of future resurrection of the dead. It may well never work, because the required neuroengineering may eventually prove too difficult or too expensive, or because future social power groups outlaw the practice, or because of many other technical or social factors. But then again, it may work if enough increased demand for such brain suspensions produces enough economies of scale and spurs enough technical and business innovation to pull it off. There is plenty of room for skepticism and variation in all the probability estimates.

But just having an affordable and plausible long shot at some type of resurrection here on Earth will, in time, affect popular belief systems and lengthen consumer time horizons. That will in turn affect risk profiles and consumption patterns, and so society will change, perhaps abruptly so. A large enough popular demand for brain suspensions would allow democracies to directly represent some of the interests from potential far-future generations, because no one would want themselves or their loved ones to revive and find a spoiled planet. Our present dead-by-one-hundred life span makes it all too easy to treat the planet like a rental car as we run up the social credit cards for unborn debtors.

The cryonics long shot lets us see our pending brain death not as the solipsistic obliteration of our world, but as the dreamless sleep that precedes a very major surgery.

SUPERINTELLIGENCE

—◦—

NICK BOSTROM

NICK BOSTROM is a philosopher at Oxford University, the author of *Anthropic Bias: Observation Selection Effects in Science and Philosophy*, and the coeditor, with Julian Savulescu, of *Human Enhancement*.

Intelligence is a big deal. Humanity owes its dominant position on Earth not to any special strength of our muscles, nor to any unusual sharpness of our teeth, but to the unique ingenuity of our brains. It is our brains that are responsible for the complex social organization and the accumulation of technical, economic, and scientific advances that, for better and worse, undergird modern civilization.

All our technological inventions, philosophical ideas, and scientific theories have gone through the birth canal of the human intellect. Arguably, human brain power is the chief rate-limiting factor in the development of human civilization.

Unlike the speed of light or the mass of the electron, human brain power is not an eternally fixed constant. Brains can be enhanced. And, in principle, machines can be made to process information as efficiently as—or more efficiently than—biological nervous systems.

There are multiple paths to greater intelligence. By "intelligence" I here refer to the panoply of cognitive capacities, including not just book-smarts but also creativity, social intuition, wisdom, and so on.

Let's look first at how we might enhance our biological brains. There are of course the traditional means: education and training, and development of better methodologies and conceptual frameworks. Also, neurological development can be improved through better infant nutrition, reduced pollution, adequate sleep and exercise, and prevention of diseases that affect the brain. We can use biotech to enhance cognitive capacity by developing pharmaceuticals that improve memory, concentration, and mental energy; or we could achieve these ends with genetic selection and genetic engineering. We can invent external aids to boost our effective intelligence—notepads, spreadsheets, visualization software.

We can also improve our collective intelligence. We can do so via norms and conventions—such as the norm against using ad hominem arguments in scientific discussions—and by improving epistemic institutions such as the scientific journal, anonymous peer review, and the patent system. We can increase humanity's joint problem-solving capacity by creating more people or by integrating a greater fraction of the world's existing population into productive endeavors, and we can develop better tools for communication and collaboration, various Internet applications being recent examples.

Each of these ways of enhancing individual and collective human intelligence holds great promise. I think they ought to be vigorously pursued. Perhaps the smartest and wisest thing the human species could do would be to work on making itself smarter and wiser. In the longer run, however, biological human brains might cease to be the predominant nexus of Earthly intelligence.

Machines will have several advantages, most obviously, faster processing speed: an artificial neuron can operate a million times faster than its biological counterpart. Machine intelligences may also have superior computational architectures and learning algorithms. These "qualitative" advantages, while harder to predict, may be even more important than the advantages in processing power and memory capacity. Furthermore, artificial intellects can be easily copied, and each new copy can—unlike humans—start life fully fledged and endowed with all the knowledge accumulated by its predecessors. Given these considerations, it is possible that one day we may be able to create "superintelligence," a general intelligence that vastly outperforms the best human brains in every significant cognitive domain.

The spectrum of approaches to creating artificial (general) intelligence ranges from completely unnatural techniques, such as those used in good old-fashioned AI, to architectures modeled more closely on the human brain. The extreme of biological imitation is whole-brain emulation or "uploading." This approach would involve creating a very detailed 3-D map of an actual brain—showing neurons, synaptic interconnections, and other relevant detail—by scanning slices of it and generating an image using computer software. Using computational models of how the basic elements operate, the whole brain could then be emulated on a sufficiently capacious computer.

The ultimate success of biology-inspired approaches seems more certain, since they can progress by piecemeal reverse-engineering of the one physical system already known to be capable of general intelligence: the brain. However, some unnatural or hybrid approach might well get there sooner.

It is difficult to predict how long it will take to develop human-level artificial general intelligence. The prospect does not seem imminent. But whether it will take a couple of decades, many decades, or centuries is probably not something that we are currently in a position to know. We should acknowledge this uncertainty by assigning some nontrivial degree of credence to each of these possibilities.

However long it takes to get from here to roughly human-level machine intelligence, the step from there to superintelligence is likely to be much quicker. In one type of scenario, the Singularity hypothesis, some sufficiently advanced and easily modifiable machine intelligence (a "seed AI") applies its wits to create a smarter version of itself. This smarter version uses its greater intelligence to improve itself even further. The process is iterative, and each cycle is faster than its predecessor. The result is an intelligence explosion. Within some very short period of time—weeks, hours—radical superintelligence is attained.

Whether abrupt and singular or more gradual and multipolar, the transition from human-level to superintelligence would be of pivotal significance. Superintelligence would be the last invention biological man would ever need to make, since, by definition, it would be much better at inventing than we are. All sorts of theoretically possible technologies could be developed quickly by superintelligence—advanced molecular manufacturing, medical nanotechnology, human enhancement technologies, uploading, weapons of all kinds, lifelike virtual realities, self-replicating space-colonizing robotic probes, and more. It would also be supereffective at creating plans and strategies, working out philosophical problems, persuading and manipulating, and much else besides.

It is an open question whether the consequences would be for the better or the worse. The potential upside is clearly enormous, but the downside includes existential risk. Humanity's future might one day depend on the initial conditions we create—in particular, on whether we successfully design the system (for example, the seed AI's goal architecture) in such a way as to make it "human-friendly," in the best possible interpretation of that term.

BECOMING ROBOTIC

———◇———

GREGORY PAUL

GREGORY PAUL is an independent researcher and the author of *Dinosaurs of the Air: the Evolution and Loss of Flight
in Dinosaurs and Birds.*

Predicting what has the potential to change everything—
really change everything—in this century is not difficult. What
I cannot know is whether I will live to see it, the data needed to
reliably calculate the span of my mind's existence being insufficient. According to the current norm, I can expect to last another
third of century. Perhaps more, if I match my grandmother's
life span; born in a Mormon frontier town the same year Butch
Cassidy, the Sundance Kid, and Etta Place sailed for Argentina,
she happily celebrated her hundredth birthday in 2001. But my
existence may exceed the natural ceiling. Modern medicine has
maximized life spans by inhibiting premature death. Sooner or
later, even that accomplishment will become passé, as advancing
technology renders death optional.

Evolution, whether biological or technological, has been
speeding up over time, as the ability to acquire, process, and
exploit information builds upon itself. Human minds adapted
to comprehend arithmetic growth tend to underestimate exponential future progress. Born two years before the Wrights' first
flight, my young grandmother never imagined she would cross
continents and oceans in near-sonic flying machines. Even out-

of-the-box thinkers did not predict the hyperexpansion of computing power over the last half century. It looks like medicine is about to undergo a similar explosion. Extracellular matrix powder derived from pig bladders can regrow a chopped-off finger with a brand new tip, complete with nail. Why not regenerate entire human arms and legs and organs?

DARPA-funded researchers Ken Muneoka, Manjong Han, and David M. Gardiner predict in *Scientific American* that we may soon be "replacing damaged and diseased body parts at will, perhaps indefinitely." Medical corporations foresee a gold mine in repairing and replacing defective organs using the cells from the victim's own body, thus avoiding the rejection problem. If assorted body parts ravaged by age can be reconstructed with tissues biologically as young and healthy as a child's, then those with the will and resources will reconstruct their entire bodies.

Even better is stopping and then reversing the very process of aging. Humans, like parrots, live exceptionally long lives, because we are genetically endowed with unusually good cellular repair mechanisms for correcting the damage created by free radicals. Given the enormous market potential, drugs are being developed to tweak genes to further upgrade the human repair system. Other pharmaceuticals are expected to mimic the life extension that appears to stem from the body's protective reaction to suppressed caloric intake. It's quite possible that middle-aged humans will be able to utilize these methods to extend their lives indefinitely. But keeping our obsolescing primate bodies and brains up and running for centuries or millennia will not be the Big Show.

The human brain and the mind it generates have not undergone a major upgrade since the Pleistocene. And they violate the basic safety rule of information processing: that it is necessary to back up data. Something more sophisticated and redundant is required. With computing power doubling every year or two, cheap personal computers should match the raw processing power of the human brain in a couple of decades, and then leave it in the dust.

If so, it should be possible to use alternative technological means to produce conscious thought. Efforts are already under way to replace damaged brain parts, such as the hippocampus, with hypercomputer implants. If and when the initial medical imperative is met, elective implants will undoubtedly be used to upgrade normal brain operations. As the fast-evolving devices improve, they will begin to outperform the original brain; it will make less and less sense to continue to do one's thinking in the old biological clunker, and formerly human minds will become entirely artificial as they move into ultrasophisticated, dispersed robot systems.

Assuming that such developments are practical, technological progress will not merely improve the human condition, but replace it. The conceit that humans in anything like their present form will be able to compete in a world of immortal superminds with unlimited intellectual capacity is naïve; there simply will not be much for people to do. Do not for a minute imagine a society of crude Terminators, or of Datas that crave to be as human as possible. Future robots will be devices of subtle sophistication and sensitivity that will expose humans as the big-brained apes we truly are. The logic predicts that most humans will choose to become robotic.

Stopping the Cyber Revolution is probably not possible. The growing knowledge base should make the production of superintelligent minds less difficult and much faster than replicating, growing, and educating human beings. Trying to ban the technology will work as well as the war on drugs has. The replacement of humanity with a more advanced system will be yet another evolutionary event—on the scale of the Cambrian Revolution or the Permian and K-T extinctions that killed off the nonavian dinosaurs and produced the advent of humans and the Industrial Age.

The scenario herein is not radical or particularly speculative. It seems so only because it has not happened yet. If the robotic civilization comes to pass, it will quickly become mundane to us. The ability of cognitive minds to adjust is endless.

Here's a pleasant secondary effect—supernaturalistic religion will evaporate, as ordinary minds become as powerful as gods. What will the cybersociety be like? I hardly have a clue. How much of this will I live to see? I'll find out.

THE SYNCHRONIZATION OF BRAINS

———— ◆ ————

JAMSHED BHARUCHA

JAMSHED BHARUCHA is a professor of psychology, provost, and senior vice president of Tufts University.

An understanding of how brains synchronize—or fail to do so—will be a game-changing scientific development.

Few behavioral forces are as strong as the delineation of in-groups and out-groups, "us" and "them." Group affiliation requires alignment, coupling, or synchronization of the brain states of members. Synchronization yields cooperative behavior, promotes group cohesion, and creates a sense of group agency greater than the sum of the individuals in the group. In the extreme, synchronization yields herding behavior. The absence of synchronization yields conflict.

People come under the grip of ideologies, emotions and moods are infectious, and memes spread rapidly through populations. Ethnic, religious, and political groups act as monolithic forces. Mobs, cults, and militias are characterized by the melding of large numbers of individuals into larger units, such that the brains of individuals operate in lockstep: a single organism controlled by a single—distributed—nervous system. Leaders who mobilize large followings have an intuitive ability to synchronize brains or to plug into systems that already are synchronized.

Herding behavior has received a great deal of attention in economics. In the recent financial bubble that eventually burst, investors and regulators were swept up by a wave of blinding optimism and overconfidence. Contrary information was discounted, and analysis from first principles ignored.

Herding behavior is prevalent in times of war. A group that perceives itself to be under attack binds together as a collective fighting unit, without questioning. When swift synchronization is critical and the stakes are high, psychological forces such as duty, loyalty, conformity, compliance—all of which promote group cohesion—come to the fore, overwhelming the rational faculties of individual brains.

Synchronization is found in many species, although the mechanisms may not be the same. Flocks of birds fly in tight formation. Fish swim in schools, and to a distant observer appear as one aggregated organism. Wolves hunt in packs. Some instances of synchronization are driven by environmental cues that regulate individual brains in the same way. For example, light cycles and seasonal cycles can entrain biorhythms of individuals who share the genetic predisposition to be regulated in this way. In other cases, the coevolution of certain behaviors together with the perception of these behaviors holds individuals together, as in the ability to both produce and recognize species-specific vocalizations.

Synchronization is mediated by communication between brains. Communicative channels include language as well as nonverbal modes such as facial expressions, gestures, tone of voice, and music. Communication across regions of an individual brain is simply a special case of a system that includes communication between brains.

Elsewhere I have argued that music serves to synchronize brain states involved in emotion, movement, and the recognition of patterns—thereby promoting group cohesion. As with tradition or ritual, what's being synchronized needn't have intrinsic utility; it may not matter what's being synchronized. The very fact of synchronization can be a powerful source of group agency.

Just around the corner is an explosion of research that regards individual brains as nodes in a system bound together by multiple channels of communication. Information technology has provided novel ways for brains to align across great distances and over time. When a song becomes a hit, millions around the world are aligned, forming a virtual unit. In the future, brain prostheses and artificial interfaces for biological systems will add to the picture.

Some clues are emerging about how brains synchronize. The hot recent discovery is the existence of mirror neurons—brain cells that respond to the actions of other individuals as if one were performing them oneself. Mirror systems are thought to generate simulations of the behavior of others in one's own brain, enabling mimicking and empathy. Other pieces of the puzzle have been accumulating for a while. Certain cases of frontal-lobe damage result in asocial behavior.

Recent work on autism has drawn attention to the mechanisms whereby individuals connect with others. The brain facilitates (sometimes in unfortunate ways) the categorization of oneself and others into in-groups and out-groups. When white participants in an MRI machine view pictures of faces, the amygdala in the left hemisphere of the brain is more strongly activated when the faces are black than when they are white. The brain has circuits specialized for the perception of faces, which convey enormous amounts of information that enable us to recognize people and gauge their emotions and intentions.

Understanding how brains synchronize to form larger systems of behavior will have vast consequences for our grasp of group dynamics, interpersonal relations, education, and politics. It will influence how we make sense of—and manage—the powerful unifying forces that constitute group behavior. For better and for worse, it will guide the development of technologies designed to interface with brains, spread knowledge, shape attitudes, elicit emotions, and stimulate action. As with all technological advances, leaders will seize on them to either improve the human condition or consolidate power.

Not all individuals are susceptible to synchronizing with others. Some reject the herd and lose out. Some chart a new course and become leaders. Being contrarian often requires enduring the psychological forces of stigma and ridicule. Understanding the conditions under which people resist will be part of the larger understanding of synchronization.

Understanding how brains synchronize—or fail to do so— will not emerge from a single new idea but rather from a complex puzzle of scientific advances woven together. What is game changing is that only recently have researchers begun to frame questions about brain function in terms not of individual brains but of how individual brains are embedded in larger social and environmental systems that drive their evolution and development. This new way of framing brain and cognitive science— together with unforeseen technological developments—promises transformational integrations of current and future knowledge about how brains interact.

THINKING SMALL: UNDERSTANDING THE BRAIN

◦

IRENE PEPPERBERG

IRENE PEPPERBERG is a research associate in psychology at Harvard University and the author of *Alex & Me: How a Scientist and a Parrot Discovered a Hidden World of Animal Intelligence—And Formed a Deep Bond in the Process.*

Knowledge of exactly how the brain works will change everything. Despite all our technical advances in brain mapping, we still do not fully understand how the human or nonhuman brain works as a complete organ, e.g., the interconnectedness of the separate areas we are currently mapping. Just as we are beginning to learn that it is not "the" gene that controls what happens in our bodies, but rather the interplay of many genes, proteins, and environmental influences that turn genes on and off, we will learn how the interplay of various neural tissues, the chemicals in our body, environmental influences, and possibly some current unknowns come together to affect how the brain works. And that will change everything.

We will, for example:

(a) ameliorate diseases in which the brain stops working properly—diseases involving cognitive deficits, such as Alzheimer's, and also those involving issues of physical control, such as Parkinson's. We will monitor just when the brain stops functioning optimally and begin interventions much earlier. Age-related senility, with its concomitant problems and societal costs, will cease to exist. If dysfunctions such as autism and schizophrenia are indeed the result of faulty interconnections among many disparate areas, we will "rewire" the appropriate systems either physically or through targeted drug intervention—and do the same for such problems as dyslexia and ADHD;

(b) understand and repair brains susceptible to addictions or to criminality based on lack of inhibitory control;

(c) use this knowledge to develop models of brain function for advanced robotics and computers, in order to design "smart" interactive systems for, among other things, space and ocean exploration, or seamless interfaces for, among other things, vision, hearing, and artificial limbs;

(d) determine ways in which human and nonhuman brains function similarly and differently; whether human and nonhuman intelligences are distinct or whether a measureable gradient exists; the extent of any overlap of function; and whether the critical issues involve modules or a constellation of interfunctioning areas that both match and are disparate. For example, we will better understand how human intelligence and language evolved and the extent to which parallel intelligence and communication evolved in nonmammalian evolutionary lines. And how they may still be evolving.

(e) attempt (maybe frighteningly) to improve on the current human brain anatomically or (more acceptably) determine what forms of teaching and training enable learning to proceed most rapidly, by enhancing appropriate connectivity and memory formation. Different types of intelligence will likely be found to be correlated with particular brain organizational patterns; thus we will identify geniuses of particular sorts more readily and cultivate their abilities.

By truly understanding brain function and harnessing it most effectively, we will affect everything else for better or worse.

CONTROLLING THE BRAIN'S PLASTICITY

LEO M. CHALUPA

LEO M. CHALUPA is vice president for research at George Washington University in Washington, DC.

In the 1967 movie *The Graduate,* a young Dustin Hoffman was advised to go into plastics, presumably because that would be the next big thing. Today, one might well advise the young person planning to pursue a degree in medicine or the biological sciences to go into brain plasticity. Our neurons are malleable throughout life, capable of being shaped by external experience and endogenous events. Recent imaging studies of single neurons have revealed that specialized parts of nerve cells, termed dendritic spines, are constantly undergoing a process of rapid expansion and retraction.

While brain cells are certainly capable of structural and functional changes throughout life, an extensive scientific literature has shown that plasticity in the nervous system is greatest early in development, which accounts for the marvelous ability of children to rapidly master various skills at different developmental stages. Toddlers have no difficulty in learning two, three, and even more languages, and most adolescents can learn to ski black-diamond slopes well before their middle-aged parents can. The critical periods underlying such learning reflect the high degree of plasticity exhibited by specific brain circuits during the first two decades of life.

In recent years, developmental neurobiologists have made considerable progress in unraveling the myriad factors underlying neuronal plasticity in the developing brain. For instance, studies have now demonstrated that it is the formation of inhibitory circuits in the cortex that causes decreased plasticity in the maturing visual system. While no single event can entirely explain brain plasticity, progress is proceeding rapidly, and I am convinced that in my lifetime we will be able to control the level of plasticity exhibited by mature neurons. Several laboratories have already discovered ways to manipulate the brain so as to make mature neurons as plastic as they are during early development. Such studies have been done using genetically engineered mice, with either a deletion or an overexpression of specific genes known to control plasticity during normal development. Moreover, drug treatments have now been found that mimic the changes observed in these mutant mice.

In essence, this means that the high degree of brain plasticity normally evident only during early development can now be made to occur throughout the life span. This is undoubtedly a game changer in the brain sciences. Imagine being able to restore the plasticity of neurons in the language centers of your brain, enabling you to learn a new language quickly and effortlessly. The restoration of neuronal plasticity would also have important clinical implications, since new connections can sprout in the developing brain, unlike in the mature brain. This technology could thus provide a powerful means to combat loss of neuronal connections, including those resulting from brain injury and various disease states.

I am optimistic that such treatments will be forthcoming in my lifetime. Indeed, a research group in Finland is carrying out the first clinical study to assess the ability of drug treatments to restore plasticity to the visual system of adult humans. If successful, this would provide a means for treating amblyopia ("lazy eye") in adults, which today can be treated only in young children whose visual cortex is still plastic.

Still, a number of problems will need to be worked out before the restoration of neuronal plasticity becomes a viable procedure. For one thing, we need to devise a means of targeting specific groups of neurons, those controlling the function whose plasticity we want to enhance. Many people might wish to have a brain able to effortlessly learn foreign languages, but few would be pleased if this new skill were accompanied by a vocabulary limited to babbling sounds, not unlike those of my granddaughter, who is beginning to learn to speak both English and Ukrainian. Targeting is everything.

NEVER-ENDING CHILDHOOD

—◦—

ALISON GOPNIK

ALISON GOPNIK is a psychologist at the University of California at Berkeley and the author of *The Philosophical Baby*.

The world is transforming from an agricultural and manufacturing economy to an information economy. This means that people will have to learn more and more. The best way to make it happen is to extend the period when we learn the most: childhood. Our new scientific understanding of neural plasticity and gene regulation, along with the global spread of schooling, will make that increasingly possible. We may remain children forever—or at least for much longer.

Humans already have a longer period of protected immaturity—a longer childhood—than any other species. Across species, a long childhood is correlated with an evolutionary strategy that depends on flexibility, intelligence, and learning. There is a developmental division of labor. Children get to learn freely about their particular environment without worrying about their own survival—caregivers look after that. Adults use what they learned as children to mate, hunt, and generally succeed as grown-ups in that environment. Children are the R&D department of the human species. We grown-ups are production and marketing. We start out as brilliantly flexible but helpless and dependent babies, great at learning everything but terrible at doing just about anything. We end up as much less flexible but much more efficient and effective adults, not so good at learning but terrific at planning and acting.

These changes reflect brain changes. Young brains are more connected, more flexible, and more plastic, but less efficient. As we get older, and experience more, our brains prune out the less-used connections and strengthen the connections that work. Recent developments in neuroscience show that this early plasticity can be maintained and even reopened in adulthood. And we've already invented the most unheralded but most powerful brain-altering technology in history: school.

For most of human history, babies and toddlers used their spectacular, freewheeling, unconstrained learning abilities to understand fundamental facts about the objects, people, and language around them—the human core curriculum. At about age six, children also began to be apprentices. Through a gradual process of imitation, guidance, and practice, they began to master the adult skills of their particular culture: hunting, cooking, navigation, child-rearing itself. Around adolescence, motivational changes associated with puberty drove children to leave the protected cocoon and act independently. By that time, their long apprenticeship had given them a new suite of executive abilities—abilities for efficient action, planning, control, and inhibition, governed by the development of prefrontal areas of the brain. By adolescence, children wanted to end their helpless status and act independently, and they had the tools to do so effectively.

School, a very recent human invention, completely alters this program. Schooling replaces apprenticeship. School lets us all continue to be brilliant but helpless babies. It lets us learn a wide variety of information flexibly and for its own sake, without any immediate payoff. School assumes that learning is more important than doing and that learning how to learn is most important of all. But school is also an extension of the period of infantile dependence—since we don't actually do anything useful in school, other people need to take care of us—all the way up to a PhD. School doesn't include the gradual mastery of specific adult skills, a mastery we once acquired in apprenticeship. Universal and extended schooling means that the period of flexible learning and dependence can continue until we are in our thirties, while independent, active mastery is increasingly delayed.

Schooling is spreading inexorably throughout the globe. A hundred years ago, hardly anyone went to school; even now, few people are schooled past adolescence. A hundred years from now, we can expect that most people will still be learning into their thirties and beyond. Moreover, the new neurological and genetic developments will give us new ways to keep the window of plasticity open. And the spread of the information economy will make genetic and neurological interventions, as well as educational and behavioral interventions, more and more attractive.

These accelerated changes have radical consequences. Schooling alone has already had a revolutionary effect on human learning: Absolute IQs have increased at an astonishing and accelerating rate (the so-called Flynn effect). Extending the period of immaturity indeed makes us much smarter and far more knowledgeable. Neurological and genetic techniques can accelerate this process even further. We all tend to assume that extending this period of flexibility and openness is a good thing—who would argue against making people smarter?

But there may be an intrinsic trade-off between flexibility and effectiveness, between the openness we require for learning and the focus we need in order to act. Childlike brains are great for learning but not so good for effective decision making or productive action. There is already some evidence that adolescents now have increasing difficulty making decisions and acting independently, and pathologies of adolescent action—such as impulsivity and anxiety—are at historical highs. Fundamental grown-up human skills we once mastered through apprenticeship—such as cooking and care giving—can't be acquired through schooling. (Think of all those neurotic new parents who have never taken care of a child and try to make up for it with parenting books.) When we are all babies forever, who will be the parents? When we're all children, who will be the grown-ups?

THE EBB OF MEMORY

KEVIN SLAVIN

KEVIN SLAVIN is cofounder and managing director of Area/Code, a company that creates cross-media games and entertainment.

In just a few years, we'll see the first generation of adults whose every breath has been drawn on the grid. A generation for whom every key moment (for example, birth) has been documented and distributed globally. Not just the key moments, of course, but also the most banal: eating pasta, missing the train, having a bad day at the office. Ski trips and puppies.

These ski trips and puppies are not simply happening, they are becoming data, building up the global database of distributed memories. They are networked digital photos—three billion on Flickr, ten billion on Facebook. They were blog posts, and now they are tweets, too (a billion in eighteen months). They are Facebook posts, Dopplr journals, Last.FM updates.

Further, more and more of our traces will be passive or semi-passive. Consider Loopt, which allows us to track ourselves and our friends with GPS (Global Positioning System). Consider voice-mail transcription bots that transcribe voice messages into searchable text in e-mail boxes, on into eternity. The next song you listen to will likely be stored in a database record somewhere. The next photo you take might well have the subject's latitude and longitude baked into the photo's metadata.

The sharp upswing in all of this record-keeping, both active and passive, is redefining one of the core elements of what it means to be human—namely, *to remember*. We are moving toward a culture that outsources this essential quality of existence to machines—a vast and distributed prosthesis. This infrastructure is up and running, but very soon we'll be living with the first generation of adults whose entire lives are embedded in it.

In 1992, the artist Thomas Bayrle wrote that one of the great mistakes of the future would be that as everything became digital we would confuse *memory* with *storage*. What's important about genuine memory, and how it differs from digital storage, is that it is imperfect: fallible and malleable. It disappears over time in a rehearsal and echo of mortality; our abilities to remember, distort, and forget are what make us who we are.

We have built the infrastructure that renders forgetting impossible. As it hardens and seeps into every element of daily life, it will make remembering impossible as well. Changing what it means to remember changes what it means to be.

There are a few people—neurological edge cases—who have perfect episodic memory, total recall. They are harbingers of the culture to come. One of them, a forty-three-year-old woman named Jill Price, was profiled in *Der Spiegel*:

> *In addition to good memories, every angry word, every mistake, every disappointment, every shock and every moment of pain goes unforgotten. Time heals no wounds for Price. "I don't look back at the past with any distance. It's more like experiencing everything over and over again, and those memories trigger exactly the same emotions in me. It's like an endless, chaotic film that can completely overpower me. And there's no stop button."*

This also describes the life of Steve Mann, a computer engineer at the University of Toronto, who has been passively recording his life through wearable computers for many years. This is an unlikely future scenario, but like any caricature it is based on human features that will be increasingly recognizable. The processing, recording, and broadcasting prefigured in Mann's work will be embedded in everyday actions, like the tweets, phonecam shots, and GPS traces we broadcast now. All of them entering into an outboard memory accessible and searchable everywhere we go.

As I write this, it is New Year's Eve. Twitter informed me today that three friends, independent of one another, were looking back at Flickr to recall what they had been doing a year ago. I would like to start the New Year being able to remember 2008 but also being able to forget it. For the next generation, each passing year will be impossible to forget and harder to remember. What will change everything is our ability to remember what everything is. Was. And wasn't.

ARTIFICIAL SELF-REPLICATING MEME MACHINES

---◦---

SUSAN BLACKMORE

SUSAN BLACKMORE is a psychologist and the author of *Consciousness: An Introduction*.

All around us, the techno-memes are proliferating and gearing up to take control. Not that they realize it; they are just selfish replicators, doing what selfish replicators do: getting copied whenever and wherever they can, regardless of the consequences. In this case, they are using us human meme machines as their first-stage copying machinery, until something better comes along. Artificial meme machines are improving all the time, and when these machines become self-replicating, that will change everything. Then they will no longer need us. Whether we live or die, or whether the planet is habitable for us or not, will be of no consequence for their further evolution.

I like to think of our planet as one in a million, or one in a trillion, of possible planets where evolution begins. This requires something (a replicator) that can be copied with variation, and selection. As Darwin realized, if more copies are made than can survive, then the survivors will pass on to the next generation of copying whatever helped them get through. This is how all design in the universe comes about.

What is not so often thought about is that one replicator can piggy-back on another by using its vehicles as copying machinery. This has happened here on Earth. The first major replicator (the only one for most of Earth's existence and still the most prevalent) is genes. Plants and animals are gene machines— physical vehicles that carry genetic information and compete to protect and propagate it. But something happened here on Earth that changed everything: One of these gene vehicles, a bipedal ape, became capable of imitation.

Imitation is a kind of copying. These apes copied actions and sounds and made new variations and combinations of old actions and sounds, and so they let loose a new replicator: memes, the cultural ideas and practices they copied. After just a few million years, the original apes were transformed, gaining enormous brains, dexterous hands, and redesigned throats and chests, in order to copy more sounds and actions more accurately. They had become meme machines.

We have no idea whether there are any other two-replicator planets out there in the universe, because their inhabitants wouldn't be able to tell us. What we do know is that our planet is now in the throes of gaining a third replicator—the step that would allow interplanetary communication.

The process began slowly and speeded up, as evolutionary processes tend to do. Marks on clay preserved verbal memes and allowed more people to see and copy them. Printing meant higher copying fidelity and more copies. Railways and roads spread the copies more widely, and people all over the planet clamored for them. Computers increased both the numbers of copies and their fidelity. The way this is usually imagined is as a process of human ingenuity creating wonderful technology as tools for human benefit and with us in control. This is a frighteningly anthropocentric way of thinking about what is happening. Look at it another way:

Printing presses, rail networks, telephones, and photocopiers were among early artificial meme machines, but they carried out only one or two of the three steps of the evolutionary algorithm. For example, books store memes and printing presses copy them, but humans still do the varying (that is, writing the books, by combining words in new ways) and the selecting (by choosing which books to buy, to read, or to reprint). Mobile phones store and transmit memes over long distances, but humans still vary and select the memes. Even on the Internet, most of the selection is still being done by humans—but this is changing fast. As we old-fashioned, squishy, living meme machines have become overwhelmed with memes, we are happily allowing search engines and other software to take over the final process of selection.

Have we inadvertently let loose a third replicator that is piggy-backing on human memes? I think we have. The information these machines copy is not human speech or actions; it is digital information competing for space in giant servers and electronic networks, copied by extremely high-fidelity electronic processes. Once all three processes of copying, varying, and selecting are done by these machines, then a new replicator has truly arrived. We might call these level-three replicators "temes" (technological memes) or "tremes" (tertiary memes). Whatever we call them, they and their copying machinery are here now. We thought we were creating clever tools for our own benefit, but in fact we were being used by blind and inevitable evolutionary processes as a steppingstone to the next level of evolution.

When memes coevolved with genes, they turned gene machines into meme machines. Temes are now turning us into teme machines. Many people work all day copying and transmitting temes. Human children learn to read very young—a wholly unnatural process that we've just got used to—and people are beginning to accept cognitive-enhancing drugs, sleep-reducing drugs, and even electronic implants to enhance their teme-handling abilities. We go on thinking we are in control, but looking at from the temes' point of view we are just willing helpers in their evolution.

So what is the step that will change everything? At the moment, temes still need us to build their machines and run the power stations, just as genes needed human bodies to copy them and provide their energy. But we humans are fragile, dim, low-quality copying machines, and we need a healthy planet with the right climate and the right food to survive. The next step is when the machines we thought we created become self-replicating. This may happen first with nanotechnology, or it may evolve from servers and large teme machines being given their own power supplies and the ability to repair themselves.

Then we would become dispensable. That really would change everything.

MALTHUSIAN INFORMATION FAMINE

---◁◦▷---

CHARLES SEIFE

CHARLES SEIFE is a professor of journalism at New York University and the author of *Decoding the Universe: How the New Science of Information Is Explaining Everything in the Cosmos, from Our Brains to Black Holes.*

For the first time, humans are within reach of a form of immortality. Just a few years ago, we had to be content with archiving a mere handful of events in our lives—storing what we could in a few faded photographs of a day at the zoo, a handful of manuscript pages, a jittery video of an anniversary, or a family legend passed down for three or four generations. All else, all of our memory and knowledge, would melt away when we died.

That era is over. It's now within your means to record, in real time, audio and video of your entire existence. A tiny camera and microphone could wirelessly transmit and store everything you see and hear for the rest of your life. It would take only a few thousand terabytes of hard-drive space to archive someone's entire audiovisual experience from cradle to grave.

Cheap digital memory has already begun to alter our society, at least on a small scale. CDs have become as quaint as LPs; now you can carry your entire music collection on a device the size of a credit card. Photographers no longer have to carry bandoliers full of film rolls. Vast databases, once confined to rooms full of spinning magnetic tapes, now wander freely about the world every time a careless government employee misplaces his laptop. Google is busy trying to snaffle up all the world's literature and convert it into a digital format—a task that, astonishingly, now has more legal hurdles than technical ones.

Much more important, though, is that vast amounts of digital memory will change the relationship humans have with information. For most of our existence, our ability to store and relay knowledge has been very limited. Every time we figured out a better way to preserve and transmit data to our peers or our descendants—as we moved from oral history to written language to the printing press to the computer age—our civilization took a great leap. Now we are reaching the point where we can archive every message, every telephone conversation, every communication between human beings anywhere on the planet. For millennia, we were starving for information to serve as raw material for ideas. Now we are about to have a surfeit.

Alas, there will be famine in the midst of all that plenty. There are some hundred million blogs, and the number is roughly doubling every year; the vast majority are unreadable. Several hundred billion e-mail messages are sent every day; most of it (current estimates run around 70 percent) is spam. There seems to be a Malthusian principle at work: Information grows exponentially, but useful information grows only linearly. Noise will drown out signal. The moment that we as a species finally have the memory to store our every thought, etch our every experience into a digital medium, it will be hard to avoid slipping into a Borgesian nightmare in which we are engulfed by our own mental refuse.

We are at the brink of a colossal change: Our knowledge is now being limited not only by our ability to gather information and to remember it, but also by our wisdom about when to ignore information—and when to forget it.

READING MINDS

—◄◦►—

KENNETH W. FORD

KENNETH W. FORD is a retired physicist and the coauthor, with John Archibald Wheeler, of *Geons, Black Holes, and Quantum Foam: A Life in Physics.*

Not in my lifetime, but someday, somewhere, some team will figure out how to read your thoughts from the signals emitted by your brain. This is not in the same league as human teleportation, which is theoretically possible but in truth fictional. Mind reading is, it seems to me, quite likely. And, as we know from hard disks and flash memories, to be able to read is to be able to write. Thoughts will be implantable.

Some will applaud the development. After all, it will aid the absentminded, enable the mute to communicate, preempt terrorism and crime, and conceivably aid psychiatry. (It will also cut down on texting and provide as reliable a staple for cartoonists as the desert island and the bed.) Some will, quite rightly, deplore it. It will be the ultimate invasion of privacy.

Game changing indeed. If we choose to play the game. Until about forty years ago, we lived in the "if it is technically feasible, it will happen" era. Now we are in the "if it is technically feasible, we can choose" era. An important moment was the decision in the United States in 1971 not to develop a supersonic transport. An American SST would hardly have been game changing, but the decision not to build it was a watershed moment in the history of technology. Of course, since then—if I may offer up my opinions—we should have said no to the International Space Station but didn't, and we should have said yes to the Superconducting Super Collider but didn't. Our skill in choosing needs refinement.

As what is technically feasible proliferates in its complexity, cost, and consequences for humankind, we should more often ask the question, "Ought we to do it?" Take mind reading. We can probably safely assume that the needed device would have to be located close to the brain being read. That would mean that choice is possible. We could let Mind Reader™, Inc., make and market it. Or we could outlaw it. Or we could hold it as an option for special circumstances (much as we now try to do with wiretapping). What we will not have to do is throw up our hands and say, "Since it can be done, it will be done." I like being able to keep some of my thoughts to myself, and I hope that my descendants will have the same option.

TRUE LIE DETECTION

———◆———

SAM HARRIS

SAM HARRIS is a neuroscientist, cofounder, with Annaka Harris, of the Reason Project, a nonprofit foundation devoted to spreading scientific knowledge and secular values in society, and the author of *Letter to a Christian Nation*.

When evaluating the social cost of deception, one must consider all of the misdeeds—marital infidelities, Ponzi schemes, premeditated murders, terrorist atrocities, genocides, and so on— that are nurtured and shored up at every turn by lies. Viewed in this wider context, deception commends itself, perhaps even above violence, as the principal enemy of human cooperation. Imagine how our world would change if, when the truth really mattered, it became impossible to lie.

The development of mind-reading technology is in its infancy, of course. But reliable lie detection will be much easier to achieve than accurate mind reading. Whether or not we ever crack the neural code, enabling us to download a person's private thoughts, memories, and perceptions, we will almost surely be able to determine to a moral certainty whether a person is representing his thoughts, memories, and perceptions honestly in conversation. Compared to many of the other hypothetical breakthroughs put forward in response to this year's *Edge* question, the development of a true lie detector would represent a very modest advance over what is currently possible through neuroimaging. Once this technology arrives, it will change (almost) everything.

The greatest transformation of our society will occur only once lie detectors become both affordable and unobtrusive. Why spirit criminal defendants and hedge-fund managers off to the lab for a disconcerting hour of brain scanning? Instead, there may come a time when every courtroom or boardroom will have the requisite technology discreetly concealed behind its wood paneling. Thereafter, civilized people would share a common presumption: that wherever important conversations are held, the truthfulness of all participants will be monitored. Well-intentioned people will happily pass between zones of obligatory candor, and these transitions will cease to be remarkable. Just as we've come to expect that many public spaces will be free of nudity, sex, loud swearing, and cigarette smoke—and now think nothing of the behavioral changes demanded of us whenever we leave the privacy of our homes—we may come to expect that certain places and occasions will require scrupulous truth telling. Most of us will no more feel deprived of the freedom to lie during a press conference or a job interview than we currently feel deprived of the freedom to remove our pants in a restaurant. Whether or not the technology works as well as we hope, the belief that it generally does work will change our culture profoundly.

In a legal context, some scholars have already begun to worry that reliable lie detection will constitute an infringement of a person's Fifth Amendment privilege against self-incrimination. But the Fifth Amendment has already succumbed to advances in our technology. The Supreme Court has ruled that defendants can be forced to provide samples of their blood, saliva, and other physical evidence that may incriminate them. In fact, the prohibition against compelled testimony appears to be a relic of a more superstitious time: It was once widely believed that lying under oath would damn a person's soul for eternity. I doubt whether even many fundamentalist Christians now imagine that an oath sworn on a courtroom Bible has such cosmic significance.

Of course, no technology is ever perfect. Once we have a proper lie detector in hand, we will suffer the caprice of its positive and negative errors; needless to say, such errors will raise real ethical and legal concerns. But some rate of error will, in the end, be judged acceptable. Remember that we currently lock people away in prison for decades—or kill them—all the while knowing that some percentage of those convicted must be innocent, whereas some percentage of those returned to our streets will be dangerous psychopaths guaranteed to reoffend. We have no choice but to rely upon our criminal justice system, despite the fact that judges and juries are poorly calibrated truth detectors, prone to error. Anything that can improve even slightly the performance of this ancient system will raise the quotient of justice in our world.

There are several reasons to doubt whether any of our current modalities of neuroimaging, such as functional magnetic resonance imaging (fMRI), will yield a practical form of mind-reading technology. It is also true that the physics of neuroimaging may grant only so much scope to human ingenuity. It is possible, therefore, that an era of cheap, covert lie detection will never dawn and we will be forced to rely on some relentlessly costly, cumbersome technology. Even so, I think it is safe to say that the time is not far off when lying about the weightiest matters will become a practical impossibility. This fact will be widely publicized, of course, and the relevant technology will be expected to be in place or accessible whenever the stakes are high. This very assurance, rather than the incessant use of these machines, will make all the difference.

RADIOTELEPATHY: DIRECT COMMUNICATION FROM BRAIN TO BRAIN

FREEMAN DYSON

FREEMAN DYSON is a theoretical physicist at the Institute for Advanced Study and the author of *The Scientist as Rebel*.

What will change everything? What game-changing scientific ideas and developments do you expect to live to see?

Since I am eighty-five, I cannot expect to see any big changes in science during my lifetime. I beg permission to change the question to make it more interesting: What game-changing scientific ideas and developments do you expect your grandchildren to see?

I assume that some of my grandchildren will be alive for the next eighty years, long enough for neurology to become the dominant game-changing science. I expect that genetics and molecular biology will be dominant for the next fifty years, and after that neurology will have its turn. Neurology will change the game of human life drastically, as soon as we develop the tools to observe and direct the activities of a human brain in detail from the outside.

The essential facts that will make detailed observation or control of a brain possible are the following: Microwave signals travel easily through brain tissue for a few centimeters. The attenuation is small enough so that signals can be transmitted from the inside and detected on the outside. Small microwave transmitters and receivers have bandwidths of the order of giga-hertz, while neurons have bandwidths of the order of kilohertz. A single microwave transmitter inside a brain has enough band-width to transmit to the outside the activity of a million neurons. A system of 10^5 tiny transmitters inside a brain, with 10^5 receivers outside, could observe in detail the activity of an entire human brain with 10^{11} neurons. A system of 10^5 transmitters outside, with 10^5 receivers inside, could control in detail the activity of 10^{11} neurons. The microwave signals could be encoded so that each of the 10^{11} neurons would be identified by the code of the signal that it transmits or receives.

These physical tools would make possible the practice of "ra-diotelepathy," the direct communication of feelings and thoughts from brain to brain. The ancient myth of telepathy induced by occult and spooky action-at-a-distance would be replaced by a prosaic kind of telepathy induced by physical tools. To make ra-diotelepathy possible, we have only to invent two new technolo-gies: first, the direct conversion of neural signals into radio signals and vice versa; and second, the placement of microscopic radio transmitters and receivers within the tissue of a living brain. I have no idea how these inventions will be achieved, but I expect them to emerge from the rapid progress of neurology before the twenty-first century is over.

It is easy to imagine radiotelepathy as a powerful instrument of social change, used for either good or evil purposes. It could be a basis for mutual understanding and peaceful cooperation of humans all over the planet. Or it could be a basis for tyrannical oppression and enforced hatred between one communal society and another. All we can say for certain is that the opportunities for human experience and understanding would be radically en-larged. A society bonded by radiotelepathy would experience hu-man life in a totally new way.

It will be our grandchildren's task to work out the rules of the game so that the effects of radiotelepathy remain constructive rather than destructive. It is not too soon for them to begin thinking about the responsibilities they will inherit. The first rule of the game, which should not be too difficult to translate into law, is that every individual should be guaranteed the ability to switch off radio communication at any time, with or without cause. When the technology of communication becomes more and more intrusive, privacy must be preserved as a basic human right.

Another set of opportunities and responsibilities will arise when radiotelepathy is extended from humans to other animal species. We will then experience directly the joy of a bird flying or a wolf pack hunting, the pain of a deer hunted or an elephant starved. We will feel in our own flesh the community of life to which we belong. I cannot help hoping that the sharing of our brains with our fellow creatures will make us better stewards of our planet.

LITTLE CHANGES MAKE THE BIGGEST DIFFERENCE

---◦◦---

BARRY C. SMITH

BARRY C. SMITH is director of the Institute of Philosophy at the University of London's School of Advanced Study and coeditor, with Crispin Wright and Cynthia Macdonald, of *Knowing Our Own Minds*.

It is tempting, when thinking about fundamental change, to look to the recent past—to the Internet and the cell phone. To lose contact with the former even temporarily can make one feel that one has been stripped of a sense, like the temporary loss of one's sight or hearing. And the ready supply of mobile-phone technology has stimulated the demand to communicate: Why be alone anywhere? You can always summon company.

Neither of these technologies is yet optimal, and either we, or they, will have to adapt. The familiar complaints are that e-mail increases our workload and cell phones put us at the end of an electronic leash. E-mail can also be an inflammatory medium, and cell phones can distract us from what is going on around us. Thus we are left uneasy with these technologies: can't live with them, can't live without them. Can future technology help—or is it we who will adapt?

Workers in AI used to dream of the talking typewriter, and this is ever closer to becoming an everyday reality. Why write e-mails when you can dictate them? Why read them when you can listen to them being read to you, perhaps in the voice patterns of their senders, while you're doing something else? Might this cut down on the inflammatory nature of e-mail exchanges?

The cell phone, our other indispensable communications device, has its own problems. We hear, unwanted, other people's conversations. We lose our inhibitions and our awareness of our surroundings in straining to capture the nuances of the other's speech—the subtle vocal signals that convey mood and meaning, many of which are simply missing in this medium. Maybe this is why people are more explicit on their cell phones, subconsciously aware that less of them comes across. Face-to-face, our attention is focused on each other's features; this multimodal experience can simultaneously provide so much. Without such cross-modal clues, we make a concentrated effort to tune in, often with dangerous consequences—as happens when drivers, even while using hands-free sets, forget what else they're doing. Could technology overcome these problems?

Here I am reminded of a huge change that occurred in the Middle Ages, when humans transformed their cognitive lives by learning to read silently. Originally, people read books out loud. Monks would whisper, of course, but the dedicated reading by so many in an enclosed space must have been a highly distracting affair. St. Ambrose amazed his fellow believers, including St. Augustine, by demonstrating that he could retain the information he found on the page without pronouncing the words. At the time, his skill was seen as a miracle, but gradually we learned to read by keeping things inside. With this simple adjustment, seemingly miraculous at the time, a great transformation of the human mind took place, and so began the age of intense private study so familiar to us now.

Will a similar transformation of the human mind come about in our time? Could we ever converse without talking out loud? If so, a quiet public space would be restored. Recently, we have seen that a monkey on a treadmill using only its mind could control—for a short time—the movements of a robot on a treadmill thousands of miles away. Here we would need something subtly different but no less astounding: a way of controlling in thought, and committing to send, the signals in the motor cortex that would normally travel to our articulators and ultimately issue in speech sounds. A device, perhaps implanted, would send the signals; another device, in receivers, would read them and stimulate similar signals in the motor cortex of the person with whom we were communicating. (Whether this could be done depends in part on whether the late Alvin Liberman's motor theory of speech perception is true, and it may well not be.) Could accent be retained? Maybe not, unless some way was found of coding separately—but usably—the information that voice conveys about the size, age, and sex of the speaker (though if you knew the speaker, you might have the illusion of hearing his voice). A breakthrough of this kind would allow us to share our thoughts efficiently and privately. Moreover, just as thinking distracts us less from our surroundings than listening attentively to sounds originating elsewhere, perhaps one could both communicate and concentrate on one's surroundings—while driving, say, or just navigating a crowd.

This won't be telepathy or the invasion of thought, since it would still depend on senders and receivers with the appropriate apparatus being willing to send to, and receive from, each other. We would still have to dial and answer. But would it come to feel as if we were exchanging thoughts directly? Perhaps. And maybe it would become the preferred way of communicating in public. I suspect that the experience of taking in the thoughts of others when reading a manuscript silently was once just as strange to medieval scholars. These are changes in experience that transform our minds, enabling us to be (notionally) in two places at once. It is these small changes in how we utilize our minds that may ultimately have the biggest effects on our lives.

NEURONALLY EXPRESSED MESSAGES

---◄◊►---

PETER SCHWARTZ

PETER SCHWARTZ is a futurist, a business strategist, co-founder of Global Business Network (a Monitor Company), and the author of *Inevitable Surprises: Thinking Ahead in a Time of Turbulence*.

It is obvious that many of the most powerful new technologies are likely to flow from biology, but one of the most game changing is likely to be neural control of devices. We are not far from being able to "jack in" to the Web. Why do I think so? Several new biological tools are converging to give us both understanding and new capabilities at the neuronal level.

The first new set of tools are the means to image the internal working of living cells, including neurons. The second are a variety of tools for precisely mapping complex biomolecular mechanics so that we can understand and manipulate neural functioning within the cell at a molecular level. And finally, the functional MRI is already giving us a systemic understanding of neural behavior.

Over the next several decades, there are likely to be other significant new biological tools that I have not foreseen that will only strengthen my argument. But the combination of these three is already likely to give us sufficient insight into how the brain works that we will be able to construct the means to reliably read the brain state and use that information to control external devices.

The first steps are far along in the pursuit of advanced prosthetic devices. We can already give the seriously injured the ability to control a prosthetic arm, and we have read out neural states that seem to express language. It is a few big steps from there to the brain's ability to reliably control such devices as cars, trucks, fighter jets, drones, machine tools, and so on. The most interesting will be the computer cursor/keyboard. It is not hard to imagine a piece of technology—say, like a Bluetooth earpiece—that would enable one to think a message and send it. This would be a form of one-way, electronically mediated telepathy.

Sending such a message to another person will be fairly easy, in the sense that control of the keyboard enables one to transmit. But we are much further from understanding the receiving process. We may never figure out how to receive directly into the brain. We may have to read the information as we do today—even if it is a neurally expressed message. There might be some advanced form of interface—for example, electronic contact lenses—although this is just a better computer screen. But neurotransmitting by itself will be sufficiently game changing and a step along the way to a radically different world of computer-mediated reality in every sense.

A NEW KIND OF MIND

KEVIN KELLY

KEVIN KELLY is an editor-at-large of *Wired* and the author
of *New Rules for the New Economy: 10 Radical Strategies for
a Connected World.*

It is hard to imagine anything that would "change everything"
as much as a cheap, powerful, ubiquitous artificial intelligence—
the kind of synthetic mind that learns and improves itself. A very
small amount of real intelligence embedded into an existing
process would boost its effectiveness to another level. We could
apply mindfulness wherever we now apply electricity. The ensu-
ing change would be hundreds of times more disruptive to our
lives than even the transforming power of electrification. We'd
use artificial intelligence the same way we've exploited previous
powers—by wasting it on seemingly silly things. Of course, we'd
plan to apply AI to tough research problems, such as curing cancer
or solving intractable math problems, but the real disruption will
come from inserting wily mindfulness into vending machines, our
shoes, books, tax returns, automobiles, e-mail, and pulse meters.

This additional intelligence need not be superhuman, or humanlike at all. In fact, the greatest benefit of an artificial intelligence would come from a mind that thought differently from
humans, since we already have plenty of those around. The game
changer is neither how smart this AI is, nor its variety, but how
ubiquitous it is. UCLA computer scientist Alan Kay quips that in

humans, perspective is worth eighty IQ points. For an artificial intelligence, ubiquity is worth eighty IQ points. A distributed AI, embedded everywhere that electricity goes, becomes ai—a low-level background intelligence that permeates the technium and through this saturation morphs it.

Ideally this "additional intelligence" should be not just cheap, but free. A free ai, like the free commons of the Web, would feed commerce and science like no other force I can imagine and would pay for itself in no time. Until recently, conventional wisdom held that supercomputers would first host this artificial mind and then perhaps we would get mini ones at home or add them to the heads of our personal robots. They would be bounded entities. We would know where our thoughts ended and theirs began.

However, the snowballing success of Google this past decade suggests that the coming ai will not be bounded inside a definable device. It will be on the Web; it will be like the Web. The more people who use the Web, the more it learns. The more it knows, the more we use it. The smarter it gets, the more money it makes, the smarter it will get, the more we will use it. The smartness of the Web is on an increasing-returns curve, self-accelerating each time someone clicks on a link or creates a link. Instead of dozens of geniuses trying to program an AI in a university laboratory, there are a billion people training the dim glimmers of intelligence arising between the quadrillion hyperlinks on the Web. Long before the computing capacity of a plug-in computer overtakes the supposed computing capacity of a human brain, the Web—encompassing all its connected computing chips—will dwarf the brain. In fact, it already has.

As more commercial life, science work, and daily play of humanity moves onto the Web, the potential and benefits of a Web ai compound. The first genuine ai will most likely not be birthed in a stand-alone supercomputer but in the superorganism of the billion CPUs known as the Web. It will be planetary in dimensions but thin, embedded, and loosely connected. Any device that touches this Web ai will share—and contribute to—its intelligence. Therefore all devices and processes will (need to) participate in this Web intelligence.

Stand-alone minds are likely to be viewed as handicapped, a penalty one might pay in order to have mobility in distant places. A truly off-the-grid ai could not learn as fast, as broadly, or as smartly as one plugged into six billion human minds, a quintillion online transistors, hundreds of exabytes of real-life data, and the self-correcting feedback loops of the entire civilization.

When this emerging ai arrives, it won't be recognized as intelligence at first. Its very ubiquity will hide it. We'll use its growing smartness for all kinds of humdrum chores, including scientific measurements and modeling, but because the smartness lives on thin bits of code spread across the globe and lacks a unified body, it will be faceless. You can reach this distributed intelligence in a million ways, through any digital screen anywhere on Earth, so it will be hard to say where it is. And because this synthetic intelligence is a combination of human intelligence (all of the past human learning, all of the current humans online) and the coveted zip of fast alien digital memory, it will be difficult to pinpoint *what* it is, as well. Is it our memory or a consensual agreement? Are we searching it, or is it searching us?

Although we will waste the Web's ai on trivial pursuits and random acts of entertainment, we'll also use its new kind of intelligence for science. Most important, an embedded ai will change how we do science. Really intelligent instruments will speed and alter our measurements; really huge sets of constant real-time data will speed and alter our model making; really smart documents will speed and alter our acceptance of when we "know" something. The scientific method is a way of knowing, but it has been based on how humans know. Once we add a new kind of intelligence into this method, it will have to know differently. At that point, everything changes.

THE AGE OF REPUTATION

------◦------

GLORIA ORRIGI

GLORIA ORRIGI is a researcher in philosophy at the Centre National de la Recherche Scientifique, Paris.

When asked about what will change our future, the most straightforward reply that comes to mind is the Internet. But how the Internet will change things that it has not already changed—what the next revolution on the Net will be—this is a harder matter. The Internet is a complex geography of information technology, networking, multimedia content, and telecommunication. This powerful alliance of technologies has provided not only a brand-new way of producing, storing, and retrieving information but also a giant network of ranking and rating systems in which information is valued as long as it has been filtered by other people.

My prediction for the Big Change is that the Information Age is being replaced by a Reputation Age, in which the reputation of an item—that is, how others value and rate it—will be the only way we have to extract information about it. This passion for ranking is a central feature of our contemporary practice of filtering information, in and out of the Net (two different examples, one inside the Net and the other outside, are www.ebay.com and the current financial crisis).

The next revolution will be a consequence of the effects of reputation on our practices of information gathering. Notice that this won't mean a world of collective ignorance, in which no one has opportunities to know something other than by relying on the judgment of someone else, in a sort of infinite chain of blind trust in which nobody seems to know anything for sure anymore. The Age of Reputation will be a new age of knowledge gathering, guided by new rules and principles. This is possible now, thanks to the tremendous potential of the social Web in aggregating individual preferences and choices to produce intelligent outcomes.

One of the main revolutions in Internet technologies has been the introduction by Google of the PageRank algorithm for retrieving information—that is, an algorithm that bases its search for relevant information on the structure of the links on the Web. Such algorithms extract the cultural information contained in each preference that users express by putting links on one page to others, with a mathematical cocktail of formulas that gives a particular weight to each of these connections. This process determines which pages are going to be in the first positions of a search result.

Fears about these tools are obviously many, because our control over the design of the algorithms—over the way the weights are assigned—is almost nonexistent. But let us imagine a new generation of search engines whose ranking procedures are generated by the aggregation of individual preferences expressed on these pages. No big calculations, no secret weights: The results of a query are organized according to the "grades" each of these pages has received by users who have visited them at least once and taken the time to rank them.

A social search engine based on the power of "soft" social computing will be able to take advantage of the reputation each site and page has accumulated, simply by the votes that users have expressed on it. The new algorithms for extracting information will exploit the power of the judgments of the many to produce their result. This softer Web, controlled more by human experiences than by complex formulas, will change our interaction with the Net, as well as our fears and hopes concerning it.

Hegel thought that universal history was made by universal judgments. Our history will be written from now on in terms of the judgments people express about matters around them, which will become more and more crucial for us in extracting information about those matters. According to the political philosopher Friedrich Hayek, civilization rests on the fact that we all benefit from being able to access knowledge we do not possess. That's exactly the kind of civilized cyberworld that will be made possible by social tools of aggregating judgments on the Web.

CRACKING OPEN THE LOCKBOX OF TALENT

HOWARD GARDNER

HOWARD GARDNER is the John H. and Elisabeth A. Hobbs Professor in Cognition and Education at the Harvard Graduate School of Education, adjunct professor of psychology at Harvard University, and the author of *Five Minds for the Future*.

What is talent? If you ask the average grade school teacher to identify her most talented student, she is likely to reject the question ("All my students are equally talented."). But of course this answer is rubbish. Anyone who has worked with numerous young people over the years knows that some catch on quickly, almost instantly, to new skills or understandings, while others must go through the same drill, with depressingly little improvement over time.

As wrongheaded as the teacher's response is the viewpoint put forward by some psychological researchers and most recently popularized in Malcolm Gladwell's *Outliers: The Story of Success*. This is the notion that there is nothing mysterious about talent, no need to crack open the lockbox: Anyone who works hard enough over a long period of time can end up at the top of her field.

But anyone who has the opportunity to observe or read about a prodigy—be it Mozart or Yo-Yo Ma in music, Tiger Woods in golf, John von Neumann in mathematics—knows that achievement is not just hard work. The differences between performance at time one and successive performances at times two, three, and four are vast, not simply the result of additional sweat. It is said that if algebra had not already existed, the philosopher and logician Saul Kripke would have invented it in elementary school; such a characterization would be ludicrous if applied to most individuals.

For the first time, it should be possible to delineate the nature of talent. This breakthrough will come about through a combination of findings from genetics (do highly talented individuals have a distinctive, recognizable genetic profile?), neuroscience (are there structural or functional neural signatures, and can these be recognized early in life?), cognitive psychology (are the mental representations of talented individuals distinctive when contrasted to those of hard workers?), and the psychology of motivation (why are talented individuals often said to have "a rage to learn, a passion to master"?).

This interdisciplinary scientific breakthrough will allow us to understand what is special about Picasso, Gauss, J. S. Mill. And it will illuminate the question of whether a talented person can achieve equally in different domains (could Mozart have been a great physicist? Could Newton have been a great musician?). Note, however, that it will *not* illuminate two other issues:

1. What makes someone original, creative? Talent and expertise are necessary but not sufficient.
2. What determines whether talents are applied to constructive or destructive ends?

These answers are likely to come from historical or cultural case studies rather than from biological or psychological science. Part of the maturity of the sciences is an appreciation of which questions are best left to other disciplinary approaches.

CULTURE

———◆———

TIMOTHY TAYLOR

TIMOTHY TAYLOR is an archaeologist at the University of Bradford, UK, and the author of *The Buried Soul: How Humans Invented Death*.

Culture changes everything, because culture contains everything—in the sense of things that can be named and so can be conceived. Ludwig Wittgenstein suggested that what cannot be said cannot be thought. He meant by this that language relies on a series of prior agreements. Such grammar has been shown by anthropologists to underpin the idea of any ongoing community—not just its language but its broader categories, its institutions, its metaphysics. And the same paradox is presented: How can anything new ever happen? If by "happen" we mean only personal and historical events, we miss the most crucial novelty: the way new things—new physical objects, devices, and techniques—insinuate themselves into our lives. They have new names that we must learn and new, revolutionary effects.

It does not always work like that. Resistance is common. Paradoxically, the creative force of culture also tries to keep everything the same. The social anthropologist Ernest Gellner said that humans, taken as a whole, present the most extensive behavioral variation of any species, whereas every particular cultural community is characterized by powerful norms. These are ways of being that, often through appeals to some apparently natural order, are not just mildly claimed as quintessentially human but rigidly enforced at a local level, in more or less public ways. Out-groups (whether of a different ethnicity, class, sexuality, or creed) or individuals who are in some way anomalous are suspect and challenging in their abnormality. Categories of special difference are typical foci for sacrifice, banishment, and ridicule, through which the in-group becomes not just the in-group but, indeed, a distinctly perceptible group, confident, refreshed, and culturally reproductive. This makes some sense; aberrance subverts the grammar of culture.

The level at which change can be tolerated varies greatly across social formations, but there is always a point beyond which things become intolerably incoherent. We may rightly label the most unprecedented behavior mad because, whatever relativization might be invoked to explain it, it is by definition strategically doomed; we seek to ignore it. Yet the routine expulsion of difference, apparently critical in the here and now, becomes maladaptive in any longer-term perspective. Clearly, it is change that has created our species' resilience and success, creating the vast inter- (not intra-) cultural diversity that Gellner noted. So how does change happen?

Major change often comes stealthily. Its revolutionary effect may often reside in the very fact that we do not recognize what it is doing to our behavior and so cannot resist it. Often we lack words to articulate resistance, as the invention is a new noun whose verbal effect lags in its wake. Such major change operates far more effectively through things than directly through people, is brought about not by the mad but rather by "mad scientists" whose inventions can be forgiven their inventors.

Unsurprisingly then, the societies that tolerate the least behavioral deviance are the most science averse. Science, in the broadest sense of effective material invention, challenges quotidian existence. The Amish (a quaint static ripple whose way of life will never uncover the simplest new technological fix for the unfolding hazards of a dynamic universe) have long recognized that material culture embodies weird inspirations challenging us, as eventual consumers, not with "copy what I do" but a far, far more subversive "try me."

Material culture is the thing that makes us human, having driven human evolution from the outset with its continually modifying power. Our species' particular dilemma is that in order to safeguard what we have, we have continually to change. The culture of things—invention and technology—is ever changing under the tide of words and routines whose role is to image fixity and agreement when, in reality, none exists. This form of change is no trivial thing; it is essential to our long-term survival—at least, to the long-term survival of anything we might be proud to call universally human.

MOLECULAR MANUFACTURING

——◇——

ED REGIS

ED REGIS is a science writer and the author of *What Is Life?: Investigating the Nature of Life in the Age of Synthetic Biology*.

Nothing has a greater potential for changing everything than the successful implementation of good old-fashioned nanotechnology.

I specify the old-fashioned version because nanotechnology is decidedly no longer what it used to be. Back in the mid–1980s, when Eric Drexler first popularized the concept in his book *Engines of Creation*, the term referred to a radical and grandiose molecular manufacturing scheme. The idea was that scientists and engineers would construct vast fleets of "assemblers"—molecular-scale programmable devices that would build objects, of practically any arbitrary size and complexity, from the molecules up. Program the assemblers to put together an SUV, a sailboat, or a spacecraft, and they'd do it—automatically and without human aid or intervention. Furthermore, they'd do it using cheap, readily available feedstock molecules as raw materials.

The idea sounds fatuous in the extreme—until you remember that objects as big and complex as whales, dinosaurs, and sumo wrestlers got built in a moderately analogous fashion. They began as minute, nanoscale structures that duplicated themselves, and whose successors then differentiated into specialized organs and other components. Those growing ranks of biological marvels did all this repeatedly, until eventually they had automatically assembled themselves into complex and functional macroscale entities. And the initial seed structures, the gametes, were not even designed, built, or programmed by scientists; they were just out there in the world, products of natural selection. But if nature can do that all by itself, then why can't machines be intelligently engineered to accomplish similar feats?

Latter-day "nanotechnology," by contrast, is nothing so imposing. In fact, the term has been co-opted, corrupted, and reduced to the point where what it refers to is essentially just small-particle chemistry. And so now we have "nanoparticles" in products ranging from motor oils to sunscreens, lipstick, car polish, and ski wax, and even a $420 "Nano Gold Energizing Cream" that its manufacturer claims transports beneficial compounds into the skin. Nanotechnology in this bastardized sense is largely a marketing gimmick, not likely to change anything very much, much less "everything."

But what if nanotechnology in the radical and grandiose sense actually became possible? What if, indeed, it became an operational reality? That would be a fundamentally transformative development, changing forever how manufacturing is done and how the world works. Imagine all our material needs being produced at trivial cost, without human labor and with no waste. No more sweatshops, no more smoke-belching factories, no more grinding workdays or long commutes. The magical molecular assemblers will do it all, permanently eliminating poverty in the process.

Then there would be the medical miracles performed by other types of molecular-scale devices—devices that would repair or rejuvenate your body's cells, killing the cancerous or other bad ones and nudging the rest toward unprecedented levels of youth, health, and durability. All without $420 bottles of face cream.

There's a downside to all this, of course, and it has nothing

to do with Michael Crichton–ish swarms of uncontrolled, preda-
tory nanobots hunting us down. Rather, it concerns the question
of what most of us will do when, newly unchained from our jobs
and blessed or cursed with longer life spans, we have oceans of
free time to kill. Free time is not a problem for the geniuses and
creators, but for the rest of us, what will occupy our idle hands?
You can play only so much golf.

But perhaps this is a problem that will never have to be
faced. Mainstream scientists generally pay little attention to
radical nanotechnology, regarding its more extravagant claims
as science-fictional. The late Richard Smalley, a Nobel prize–
winning chemist, made a cottage industry out of arguing that
insurmountable technical difficulties at the chemical bonding
level would keep radical nanotechnology perpetually in the pipe-
dream stage. Nobody yet knows whether he was right. Some may
hope he was. Maybe changing *everything* is not as attractive an
idea as it first seems.

RESIZING OURSELVES

———◁○▷———

DOMINIQUE GONZALEZ-FOERSTER

DOMINIQUE GONZALEZ-FOERSTER is an artist.

Following the nano and miniaturization trend occurring in many fields, from tapas to cameras, including surgery, vegetables, cars, computers. . . .

Let's imagine that the fantastic Gulliver iconography, with its cohabitation of tiny and giant people, could also have some visionary quality. Let's think that the 1957 film *The Incredible Shrinking Man* can be more than science fiction, and let's imagine a worldwide collective decision to genetically miniaturize future generations in order to reduce human needs and increase space and resources on the blue planet.

There would be a strange Gulliver-like period of transition, where giants would still live with the smaller generations, but in the long run the planet might look very different and the change of scale in relation to animals, plants, and landscapes could generate completely new ideas, perceptions, representations, and ideas.

THE ACTUAL, THE POSSIBLE, AND THE UNIMAGINABLE

---◇---

MARC D. HAUSER

MARC D. HAUSER is a psychologist and biologist at Harvard University and the author of *Moral Minds: How Nature Designed Our Universal Sense of Right and Wrong.*

Science fiction writers traffic in possible worlds. What if, as in the Hollywood blockbuster *Minority Report*, we could read people's intentions before they acted and thus prevent violence? An intentionality detector would be a terrific device to have, but talk about ethical nightmares. If you ever worried about Big Brother tapping your phone lines, how about tapping your neural lines?

What about aliens from another planet? What will they look like? How do they reproduce? How do they solve problems? If you want to find out, just go back and watch reruns of *Star Trek* or get out the popcorn and watch *Men in Black, War of the Worlds, The Thing, Signs,* or *The Blob*.

But here's the rub on science fiction: It's all basically the same stuff, one gimmick with a small twist. Look at all the aliens in these movies. They are almost always the same—a bit wispy, often with oversized heads and see-through body parts, and with awesome powers. And surprisingly, this is how it has been for a hundred or so years of motion pictures, even though our technologies have greatly expanded the range of special effects. Why the lack of creativity? Why such a poverty of the imagination?

The answer reveals a deep fact about our biology and the biology of all other organisms. The brain, as a physical device, evolved to process information and make predictions about the future. Though the generative capacity of the brain, especially the human brain, is spectacular, providing us with a system for massive creativity, it is also highly constrained. The constraints arise both from the physics of brain operation and the requirements of learnability.

These constraints establish what we and other organisms have achieved (the actual) and what they could, in the future and with the right conditions, potentially achieve (the possible). Where things get interesting is in thinking about the unimaginable. *Poof!*

But there is a different way of thinking about this problem—a way that takes advantage of exciting new developments in molecular biology, evolutionary developmental biology, morphology, neurobiology, and linguistics. In a nutshell, for the first time we have a science that enables us to understand the actual, the possible, *and* the unimaginable—a landscape that will forever change our understanding of what it means to be human, including how we arrived at our current point in evolution and where we might end up in ten, or ten million, years.

To illustrate, consider a simple example from the field of theoretical morphology, a discipline that aims to map out the space of possible morphologies and in so doing reveals not only why some parts of this space were never explored but also why they never could be explored. The example concerns an extinct group of animals called the ammonoids, swimming cephalopod mollusks with a shell that spirals out from the center before opening up.

In looking at the structure of their shells—the ones that actually evolved, that is—we see two relevant dimensions that account for the observed variations: the rate at which the spiral spirals out and the distance between the center of this spiral and the opening. If you plot the actual ammonoid species on a graph that includes spiral rate and distance to the opening, you see a density of animals in a few areas and then some gaps. The occupied spaces in this map show what actually evolved, whereas the vacant spaces suggest either possible (not yet evolved) or impossible morphologies.

Of great interest in this line of research is the cause of the impossible. Why, that is, have certain species never taken over a particular swath of morphological turf? What is it about this space that leaves it vacant? Skipping many details, some of the causes are intrinsic to the organisms (for example, no genetic material or developmental programs for building wheels instead of legs) and some are extrinsic (for example, circles represent an impossible geometry, or natural habitats would never support wheels).

What is exciting about these ideas is that they have a family resemblance to those that Noam Chomsky mapped out more than fifty years ago in linguistics. That is, the biology that allows us to acquire a range of possible languages also puts constraints on this system, leaving in its wake a space of impossible languages—those that could never be acquired or, if acquired, would never remain stable. And the same moves can be translated into other domains of cultural expression, including music, morality, and mathematics. Are there musical scores that no one, not even John Cage, could dream up because the mind can't fathom certain frequencies and temporal arrangements? Are there evolvable moral systems that we will never see because our current social systems and environments make them toxic to our moral sensibilities? Regardless of how these questions are resolved, they open up new research opportunities, using methods that are only now being refined.

Today we can extend the range of the possible, invading the terra incognita of the impossible. Thanks to work by neuroscientists such as Evan Balaban, we now know that we can combine the brain parts of different animals to create chimeras. For example, we can replace part of a chicken's brain with the anatomically similar part from a quail's brain, and when the young chick develops, it moves its head like a quail but calls like a chicken. Functionally, we have allowed the chicken to invade an empty space of behavior, something unimaginable—to a chicken, that is.

Now let your imagination run wild. What would a chimpanzee do with the generative machinery that a human has when running computations in language, mathematics, and music? Could it imagine the previously unimaginable? What if we gave a genius like Einstein the key components that made Bach a different kind of genius? Could Einstein now imagine different dimensions of musicality? These same neural manipulations are now possible at the genetic level. Genetic engineering allows us to insert genes from one species into another or manipulate the expressive range of a gene, jazzing it up or turning it off.

This revolutionary science is here, and it will forever change how we think. It will change what is possible, potentially remove what is possible but deleterious, and open our eyes to the previously impossible.

COMPUTING THE EMBRYO

---◦---

LEWIS WOLPERT

LEWIS WOLPERT is a professor of biology at University College, London, and the author of *How We Live and Why We Die: The Secret Lives of Cells.*

We know much about the mechanisms involved in the development of embryos. But given the genome of the egg, we cannot predict the way the embryo will develop. This will require an enormous computation, in which all the many thousands of components, particularly proteins, are involved and so the behavior of every cell will be known. We would, given a fertilized human egg, then be able to have a picture of all the details of the newborn baby, including any abnormalities. We would also be able to program the egg to develop into any shape we desire. The time will come when this is possible.

HOMO EVOLUTIS

—◇—

JUAN ENRIQUEZ

JUAN ENRIQUEZ is the CEO of Biotechonomy, founding director of the Harvard Business School's Life Sciences Project, and the author of *The Untied States of America: Polarization, Fracturing, and Our Future.*

Speciation is coming. Fast. We keep forgetting that we are but one of several hominids that have walked the Earth (*erectus, habilis, neanderthalis, heidelbergensis, ergaster, australopithecus*). We keep thinking we are the one and only, the special. But we easily could not have been a dominant species. Or even still be a species. We blissfully ignore the fact that we came within about two thousand specimens of going extinct (which is why all human DNA is virtually identical).

There is not much evidence historically that we are the be-all and end-all or that we will remain the dominant species. The fossil history of the planet reveals at least six mass extinctions. In each cycle, most life was toast, as DNA/RNA hit a reboot key. New species emerged to adapt to new conditions. An asteroid hits? Gone are oceans of slime. The globe freezes to the equator? Microbes dominate. The atmosphere fills with poisonous oxygen? No worries—life eventually blurts out obnoxious mammals.

Unless we believe that we have now stabilized all planetary and galactic variables, these cycles of growth and extinction will continue. Ninety-nine percent of species, including all other hominids, have gone extinct. Often this has happened over long periods of time. What is interesting today, two hundred years after Darwin's birth, is that we are taking direct and deliberate control over the evolution of many, many species, including ourselves. So the single biggest game changer will likely be the beginning of human speciation. We will get glimpses of it in our lifetime. Our grandchildren will likely live it.

There are at least three parallel tracks on which this change is running toward us. The easiest to see and comprehend is taking place among the "handicapped." As we build better prostheses, we begin to see equality. Legless Oscar Pistorius attempting to put aside the Special Olympics and run against able-bodied Olympians is but one example. In Beijing, he came very close but did not meet the qualifying times. However, as materials science, engineering, and design advance, by the next Olympics he and his disciples will be competitive. And one Olympics after that, the "handicapped" could be unbeatable.

It's not just limbs. What started out as large cones for the hard of hearing eventually became pesky, malfunctioning hearing aids. Then came discreet, effective, miniaturized buds. Now internal cochlear implants allow the deaf to hear. But unlike natural evolution, which requires centuries, digital technologies double in power and halve in price every few months. Soon those with implants will hear as well as we do, and a few months after that, their hearing may be more acute than ours. Likely the devices will span a broad and adjustable tonal range, including that of such species as dogs, bats, and dolphins. Wearers will adapt to various environments at will. Perhaps musicians with natural hearing will file antidiscrimination lawsuits because they were not hired by symphony orchestras.

Speciation does not have to be mechanical. There is a parallel, fast-moving track in stem cell and tissue engineering. While the global economy melted down last year, a series of extraordinary discoveries opened interesting options that will be remembered far longer than the drooping NASDAQ index. Laboratories in Japan and Wisconsin rebooted skin cells and turned them into stem cells. We are now closer to a point where any cell in our body can be rebooted back to its original factory settings (pluripotent stem cells) and can rebuild any part of our body. At the same time, a Harvard team stripped a mouse heart of all its cells, leaving only cartilage. The cartilage was covered in mouse stem cells, which self-organized into a beating heart. A Wake Forest group regrew human bladders and implanted them into accident and cancer victims. By year's end, a European team had taken a trachea from a dead donor, removed the cells and covered the sinew with bone-marrow cells taken from a patient dying of tuberculosis. These cells self-organized and regrew a fully functional trachea, which was implanted into the patient. There was no need for immunosuppressants; her body recognized the cells covering the new organ as her own. This is but one demonstration that treatments for the sick can quickly expand into the population, through elective procedures. The global proliferation of plastic surgery shows that many are willing to undergo great expense, pain, and inconvenience in order to enhance their bodies. Between 1996 and 2002, elective cosmetic surgery increased 297 percent and minimally invasive procedures increased 4,146 percent. As artificial limbs, eyes, ears, and cartilage begin to provide significant advantages, procedures developed to enhance the quality of life for the handicapped may become common.

After the daughter of one of my friends tore her tendons horseback riding, doctors told her that they could harvest parts of her own tendons and hamstrings to rebuild her leg. Because she was so young, the crippling procedure would have to be repeated three times as her body grew. But because her parents knew tissue engineers who were growing tendons in a lab, she became instead one of the first recipients of a procedure that allows natural growth and no harvesting. Today she is a successful ski racer, and her coach, having determined that her "damaged" knee is far stronger, has asked whether the same procedure could be done on the other one.

As we regrow or reengineer more body parts, we will likely significantly increase average life span and run into a third track of speciation. Those with access to Google already have an extraordinary evolutionary advantage over the digitally illiterate. During the next decade, we will be able to store everything we have seen, read, and heard in our lifetime. The question is, can we re-upload and upgrade this data as the basic storage organ deteriorates? And can we enhance this organ's cognitive capacity, either internally or externally? MIT has already brought together many of those interested in cognition—neuroscientists, surgeons, radiologists, psychologists, psychiatrists, computer scientists—to begin to understand this black box. Rebooting other body parts will likely be easier than rebooting the brain, so this will be the slowest track—but over the long term the one with the greatest speciation potential.

Speciation will not be a deliberate, programmed event; instead, it will involve an ever-faster accumulation of small, useful improvements that will eventually turn *Homo sapiens* into a new hominid. We will begin to see evidence of this partly mechanical, partly regrown creature who is rapidly driving its own evolution. As the branches of the tree of life, and of hominids, continue to spread, many of our grandchildren will likely engineer themselves into what we would consider a new species, one with extraordinary capabilities: *Homo evolutis*.

THE OPEN UNIVERSE

—◁◦▷—

STUART KAUFFMAN

STUART KAUFFMAN is the director of the Institute for
Biocomplexity and Informatics at the University of Calgary
and the author of *Reinventing the Sacred: A New View of
Science, Reason, and Religion*.

John Brockman's question is dramatic: What will change
everything? Of course, no one knows. But the fact that no one
knows may be the feature of our lives and the universe that does
change everything. Reductionism has reigned as the dominant
worldview in Western society for three hundred and fifty years.
Physicist Steven Weinberg states that when science shall have
been done, all the explanatory arrows will point downward, from
societies, to people, to organs, to cells, to biochemistry, to chemistry, and ultimately to physics and the final theory.

I think he is wrong: The evolution of the biosphere, the
economy, our human culture, and perhaps aspects of the abiotic
world, stand partially free of physical law and are not entailed by
fundamental physics. The universe is open.

Many physicists now doubt the adequacy of reductionism, including Philip Anderson and Robert Laughlin. Laughlin argues
for laws of organization that need not derive from the fundamental laws of physics. I give one example: Consider a sufficiently
diverse collection of molecular species—such as peptides, RNA,
or small molecules—that can undergo reactions and are also

candidates to catalyze those very reactions. It can be shown analytically that at a sufficient diversity of molecular species and reactions, so many of these reactions are expected to be catalyzed by members of the system that a giant catalyzed reaction network arises that is collectively autocatalytic. It reproduces itself.

The central point about the autocatalytic set theory is that it is a mathematical theory, not reducible to the laws of physics, even if any specific instantiation of it requires actual physical "stuff." It is a law of organization that may play a role in the origin of life.

Consider next the number of proteins with two hundred amino acids: twenty to the two-hundredth power. If the 10^{80} particles in the known universe were doing nothing but making proteins two hundred amino acids long on the Planck timescale, and if the universe is 13.7 billion years old, it would require 10^{39} lifetimes of the universe to make all possible proteins of that length just once. But this means that above the level of atoms, the universe is on a unique trajectory. It is vastly nonergodic. Then we will never be able to make all the complex molecules, organs, organisms, or social systems. In this second sense, the universe is indefinitely open "upward" in complexity.

Consider the human heart, which evolved in the nonergodic universe. I claim that the physicist can neither deduce nor simulate the evolutionary becoming of the heart. Simulation, given all the quantum throws of the dice (for example, the mutation of genes by cosmic rays), seems out of the question. And if such infinitely or vastly many simulations were to be carried out, there would be no way to determine which one had captured the evolution of this biosphere.

Suppose we asked Darwin what the function of the heart is. "Pumping blood" is his brief reply. But there is more. Darwin noted that features of an organism that are of no selective use in the current environment might be selected in a different environment. These are called Darwinian "preadaptations," or "exaptations." Here is an example: Some fish have swim bladders, partially filled with air and partially with water, that adjust neutral buoyancy in the water column. They arose from lungfish. Water got into the lungs of some fish, and now there was a sac partially filled with air, partially filled with water, poised to become a swim bladder. Three questions arise: Did a new function arise in the biosphere? Yes, neutral buoyancy in the water column. Did it have cascading consequences for the evolution of the biosphere? Yes, new species, proteins, and so forth.

Now comes the essential third question: Do you think you could predict all the possible Darwinian exaptations of all organisms alive now, or just for humans? We all seem to agree that the answer is a clear No, we cannot predict the possible exaptations. As noted in the first paragraph, *we really do not know what will happen.* Part of the problem seems to be that we cannot prespecify all possible selective environments. How would we know we had succeeded? Nor can we prespecify the feature(s) of one or several organisms which might become exaptations.

Then we can make no probability statement about such preadaptations: We do not know the space of possibilities—the sample space—so we can construct no probability measure.

Can we have a natural law that describes the evolution of the swim bladder? If a natural law is a compact description available beforehand, the answer, again, seems a clear No. *But in that case, it is not true that the unfolding of the universe is entirely describable by natural law.* This contradicts our views since Descartes, Galileo, and Newton. The unfolding of the universe seems to be partially lawless. In its place is a radically creative becoming.

Let me point to the Adjacent Possible of the biosphere. Once there were lungfish, swim bladders were in the Adjacent Possible of the biosphere. Before there were multicelled organisms, the swim bladder was *not* in the Adjacent Possible of the biosphere. Something wonderful is happening right in front of us: When the swim bladder arose, it was of selective advantage *in its con-*

text. It changed what was Actual in the biosphere, which in turn created a new Adjacent Possible of the biosphere. The biosphere self-consistently coconstructs itself into its ever-changing, unstatable Adjacent Possible.

If the becoming of the swim bladder is partially lawless, it certainly is not entailed by the fundamental laws of physics, so cannot be deduced from physics. *Then its existence in the nonergodic universe requires an explanation that cannot be had by that missing entailment. The universe is open.*

Part of the explanation rests in the fact that life seems to be evolving ever more positive-sum games. As organismic diversity increases and the "features" per organism increase, there are more ways for selection to select for mutualisms that become the conditions of joint existence in the universe. The hummingbird, sticking her beak in the flower for nectar, rubs pollen off the flower, flies to a new flower for nectar, and the pollen rubs off onto the stamen of the next flower, pollinating that flower. But these mutualistic features are the very conditions of one another's existence in the open universe. The biosphere is rife with mutualisms. In biologist Scott Gilbert's fine phrase, these are codependent origination. In this open universe, beyond entailment by fundamental physics, we have partial lawlessness, ceaseless creativity, and forever codependent origination that changes the Actual and the ever-new Adjacent Possible we ceaselessly self-consistently coconstruct. Moreover, the way this unfolds is neither fully lawful nor is it random. We need to reenvision ourselves and the universe.

LIVING TO A HUNDRED AND FIFTY

———◦———

GREGORY BENFORD

GREGORY BENFORD is a novelist, cofounder and chairman of Genescient, and the author of *The Sunborn*.

I expect to see this happen, because I'll be living longer. Maybe even to a hundred and fifty, about thirty more years than any human is known to have lived. I expect this because I've worked on it, seen the consequences of genomics when applied to the complex problem of our aging.

Since Aristotle, many scientists and even some physicians (who should know better) thought that aging arose from a few mechanisms that make our bodies deteriorate. The genomic revolution of the last decade now promises a true twenty-first-century path to extending longevity: Follow the pathways. Genomics now reveals what physicians intuited: the staggering complexity of aging pathophysiology among real clinical patients. We can't solve the "aging problem" using the standard research methods of cell biology, despite the great success such methods had with some other medical problems.

Aging is not a process of deterioration actively built by natural selection. Instead, it arises from a lack of such natural selection in later adulthood. Not understanding this is the reason for the perennial failure to explain or control aging and the chronic diseases underlying it. Aging comes from multiple genetic deficiencies, not a single biochemical problem.

But now we have genomics to reveal all the genes in an organism. More, we can monitor how each and every one of them is expressed in our bodies. Genomics, working with geriatric pathology, now unveils the intricate problems of co-ordination among aging organ systems. Population genetics illuminates aging's cause and so, soon enough, its control. Aging arises from interconnected complexity hundreds of times greater than that envisioned by cell biologists before the late 1990s.

The many-headed monster of aging can't be stopped by any vaccine or by supplying a single missing enzyme. There are no "master regulatory" genes or avenues of accumulating damage. Instead, there are complex pathways that inevitably trade current performance for long-term decay. Eventually that evolutionary strategy catches up with us.

So the aging riddle is inherently genomic in scale. There is no biochemical or cellular imperative to aging; it arises from side effects of evolution, through natural selection. But this also means that we can attack it by using directed evolution.

Michael Rose at the University of California, Irvine, has produced "Methuselah flies" that live over four times longer than control flies in the lab. He did this by not allowing their eggs to hatch until half are dead, for hundreds of generations. Methuselah flies are more robust, not less, and so resist stress.

Methuselah flies' genomics shows us densely overlapping pathways. Directed evolution uses these to enhance longevity. Since flies have about three-quarters of their genes in common with us, this tells us much about our own pathways. We now know many of these pathways and can enhance their resistance to the many disorders of aging.

By finding substances that can enhance the action of those pathways, we have a twenty-first-century approach to aging. Such research is rapidly ongoing in private companies, including one I cofounded three years ago. The field is moving fast. The genomic revolution makes the use of multipathway treatments to offset aging inevitable.

Knowledge comes first, then its use. Science yields engineering. Already there seems no fundamental reason why we cannot live to a hundred and fifty or longer. After all, nature has done quite well on her own. We know of a forty-eight-hundred-year-old bristlecone pine, a four-hundred-year-old clam—plus whales, a tortoise, and koi over two hundred years old—all without technology. These organisms use pathways we share and can now understand.

It will take decades to find the many ways of acting on the longevity genes we already know about. Nature spent several billion years developing these pathways; we must plumb them with smart modern tools. The technology emerging now acts on these basic pathways to immediately affect all types of organs. Traditionally, medicine focuses on disease by isolating and studying organs. Fair enough, for then. Now it is better to focus on entire organisms. Only genomics can do this; it looks at the entire picture.

Quite soon, simple pills containing designer supplements will target our most common disorders—cardiovascular, diabetic, neurological. Beyond that, the era of affordable personal genomics makes possible designer supplements, now called nutrigenomics. Tailored to each personal genome, these can enforce the repair mechanisms and augmentations that nature provided to the genomically fortunate.

So . . . what if it works?

The prospect of steadily extending our life spans terrifies some governments. These will yield, over time, to pressures to let us work longer—certainly far beyond the sixty-five years imposed by most European Union countries. Slowly it will dawn on us that vibrant old age is a boon, not a curse.

Living to a hundred and fifty ensures that you take the long view. You're going to live in a future ecology, so better be sure it's livable. You'll need long-term investments, so think long term. Social problems will belong to you, not to some distant others, because problems evolve and you'll be around to see them.

Rather than isolating people, "old age" will lead to social growth. With robust health to go with longer lives, the older will become more socially responsible, bringing both experience and steady energy to bear. We need fear no senioropolis of caution and withdrawal. Once society realizes that people who get educated in twenty years can use that education for another century or so, working well beyond the age of one hundred, the twentieth-century social agenda vanishes. Nobody will retire at sixty-five. People will switch careers, try out their dreams, perhaps find new mates and passions. We will see that experience can damp the ardent passions of glib youth, if it has a healthy body to work through. That future will be more mature, and richer for it.

All this social promise emerges from the genomic revolution. The twenty-first century has scarcely begun, and already it looks as though most who welcomed it in will see it out—happily, after a good swim in the morning and a vigorous party that night to welcome in the twenty-second. The first person to live to a hundred and fifty may be reading this right now.

MASTERING DEATH

---◇---

MARCELO GLEISER

MARCELO GLEISER is the Appleton Professor of Natural Philosophy at Dartmouth College and the author of *The Prophet and the Astronomer: Apocalyptic Science and the End of the World.*

There is no question more fundamental to us than our mortality. We die and we know it. It is a terrifying, inexorable truth, one of the few absolute truths we can count on. Other noteworthy absolute truths tend to be mathematical, such as $2 + 2 = 4$. Nothing horrified the French philosopher and mathematician Blaise Pascal more than "the silence of infinitely open spaces," the nothingness that surrounds the end of time and our ignorance of it.

For death is the end of time, the end of experience. Even if you are religious and believe in an afterlife, things certainly are different then: Either you exist in a timeless Paradise (or Hell) or as some reincarnate soul. If you are not religious, death is the end of consciousness. And with consciousness goes the end of tasting a good meal, reading a good book, watching a beautiful sunset, having sex, loving someone. Pretty grim in either case.

We exist only as long as people remember us. I think of my great-grandparents in nineteenth-century Ukraine. Who were they? No writings, no photos, nothing. Just their genes remain, diluted, in our current generation.

What to do? We spread our genes, write books and essays,

prove theorems, invent family recipes, compose poems and symphonies, paint and sculpt—anything to create some sort of permanence, something to defy oblivion. Can modern science do better? Can we contemplate a future in which we will we control mortality? I am doubtless being far too optimistic in considering this a possibility, but the temptation to speculate is too great.

Let's say I'll live to the age of a hundred and one, like Irving Berlin; in that case, I still have half my life ahead of me. I can think of two ways in which mortality might be tamed: one at the cellular level and the other through an integration of the body with genetics, the cognitive sciences, and cybertechnology. I'm sure there are others. But first, let me make clear that, at least according to current science, mortality can never be completely overridden. Speculation aside, modern physics forbids time travel to the past. Unfortunately, we can't just jump into a time machine to relive our youth over and over again. (Sounds a bit horrifying, actually.) Causality is an unforgiving mistress. Also, unless you are a vampire (and there have been times when I wished I were one) and thus beyond the laws of physics, you can't really escape the second law of thermodynamics: Even an open system like the human body, able to interact with its environment and absorb nutrients and energy from it, will slowly deteriorate. In time, we burn too much oxygen. We live, and we rust. Herein life's cruel compromise: We need to eat to stay alive, but by eating we slowly kill ourselves. At the cellular level, the mitochondria are the little engines that convert food into energy. Starving cells live longer. Apparently, proteins from the sirtuin (Silent Information Regulator 2) family contribute to this process, interfering with normal apoptosis, the cellular self-destruction program.

Could the right dose of sirtuin, or something else, significantly slow down aging in humans? Maybe in a few decades. Genetic action may also interfere with the usual mitochondrial respiration: Reduced expression of the mclk1 gene has been shown to slow down aging in mice. Something similar happens in *Caenorhabditis elegans* worms. These results suggest that the same molecular mechanism for aging occurs throughout the animal kingdom.

We can speculate that by, say, 2040 a combination of these two mechanisms will allow scientists to unlock the secrets of cellular aging. It's not the elixir of life alchemists dreamt of, but the average life span could possibly be increased to a hundred and twenty-five years or even longer, a significant jump from the current U.S. average of seventy-seven years or so. Of course, this would create a terrible burden on, among other things, our Social Security system—but perhaps retirement age by then would be around a hundred.

A second possibility is more daring and probably much less likely to become a reality within my next fifty or so years of life. Combine human cloning with a mechanism to store all our memories in a giant database. Inject the clone of a certain age with the corresponding memories. Voilà! Will this clone be you? No one really knows. Certainly, just the clone without the memories won't do. We are what we remember.

To keep on living with the same identity, we must keep on remembering. Unless, of course, you don't like yourself and want to forget the past. So, assuming that such a tremendous technological jump is even feasible, we could migrate to a new copy of ourselves when the current one gets old and decrepit. Some of my colleagues are betting such technologies will become available within the century.

Although I'm an optimist by nature, I seriously doubt it. I probably will never know, and my colleagues won't, either. However, there is no question that controlling death is the ultimate human dream, the one thing that can change everything else. I leave the deeply transforming social and ethical upheaval this would cause to another essay. Meanwhile, I take advice from Mary Shelley's *Frankenstein*. Perhaps there are advances we are truly unprepared for.

NO MORE TIME DECAY

———◇———

EMANUEL DERMAN

EMANUEL DERMAN is a professor of financial engineering at Columbia University, a principal of Prisma Capital
Partners, former head of Quantitative Risk Strategies Group
at Goldman Sachs, and the author of *My Life as a Quant:
Reflections on Physics and Finance*.

The biggest game changer looming in your future, if not
mine, is life prolongation. It works for mice and worms, and
surely one of these days it will work for the rest of us.

The current price for life prolongation seems to be semi-
starvation; the people who try it wear loose clothes to hide their
ribs and intentions. There's something desperate and shameful
about starving yourself in order to live longer. But right now,
biologists are tinkering with resveratrol and sirtuins, trying to
get you the benefit of life prolongation without cutting back on
calories.

Life and love get their edge from the possibility of their ending. What will life be like when we live forever? Nothing will be
the same.

The study of financial options shows that there is no free
lunch. What you lose on the swings, you gain on the round-
abouts. If you want optionality, you have to pay a price, and part
of that price is that the value of your option erodes every day.
That's time decay. If you achieve a world where nothing fades

away with time anymore, it will be because there's nothing to fade away.

No one dies. No one gets older. No one gets sick. You can't tell how old someone is by looking at them or touching them. No May-September romances. No room for new people. Everyone's an American car in Havana, endlessly repaired and maintained long after its original manufacturer is defunct. No breeding. No one born. No more evolution. No sex. No need to hurry. No need to console anyone. If you want something done, give it to a busy man, but no one need be busy when you have forever. Life without death changes absolutely everything.

If everyone is an extended LP, the turntable has to turn very slowly.

Who's going to do the real work, then? Chosen people who will volunteer or be volunteered to be mortal.

"If we want things to stay as they are, things will have to change." (Giuseppe di Lampedusa in *The Leopard*)

WEST ANTARCTICA AND SEVEN OTHER SLEEPING GIANTS

———◇———

LAURENCE C. SMITH

LAURENCE C. SMITH is a professor of geography and earth and space sciences at UCLA.

In the classic English fairy tale *Jack and the Beanstalk*, the intrepid protagonist risks being devoured on sight in order to repeatedly raid the home of a flesh-eating giant for gold. All goes well until the snoring giant awakens and gives furious chase. But Jack beats him back down the magic beanstalk and chops it down with an axe, toppling the descending cannibal to his death. Jack thus wins back his life plus substantial economic profit from his spoils.

Industrialized society has also reaped enormous economic and social benefit from fossil fuels, so far without rousing any giants. But as geoscientists, my colleagues and I devote much of our time to worrying about whether giants might be slumbering in Earth's climate system. We used to think climate worked like a dial—slow to heat up and slow to cool down—but we've since learned it can also act like a switch. Twenty years ago, anyone who hypothesized an abrupt, show-stopping event—a centuries-long plunge in air temperature, say, or the sudden die-off of forests—would have been laughed at. But today an immense body of empirical and theoretical research tells us that sudden awakenings are dismayingly common in climate behavior.

Ancient records preserved in tree rings, sediments, glacial

ice layers, cave stalactites, and other natural archives tell us that for much of the past ten thousand years—during which time our modern agricultural society evolved—our climate was remarkably stable. Before then, it was capable of wild fluctuations, even leaping 18° F in ten years. That's as if the average temperature in Minneapolis were to warm to that of San Diego in the same amount of time.

Even during the relative calm of recent centuries, we find sudden lurches that exceed anything in modern memory. Tree rings tell us that in the past thousand years the western United States has seen three droughts at least as bad as the Dust Bowl but lasting three to seven times longer. Two of them may have helped collapse the past societies of the Anasazi and Fremont people.

The mechanisms behind such lurches are complex but decipherable. Many are related to shifting ocean currents that move regions of warm or cool seawater in quasi-predictable ways. The El Niño/La Niña phenomenon, which redirects rainfall patterns around the globe, is one well-known example. Another major player is the Atlantic thermohaline circulation, or THC—a massive, density-driven "heat conveyor belt" that carries tropical warmth northward via the Gulf Stream. The THC is what keeps Europe relatively balmy despite being as far north as some of Canada's best polar-bear habitat. If the THC were to weaken or halt, the eastern United States and Europe would become something like Alaska. While oversensationalized by the 2004 film *The Day After Tomorrow* and a scary 2003 Pentagon document imagining famines, refugees, and wars, a THC shutdown nonetheless remains an unlikely but plausible threat. It is the original sleeping giant of my field.

Unfortunately, we are discovering more giants that are probably lighter sleepers than the THC. Seven others—all of them potential game changers—are now under scrutiny:

1. the disappearance of summer sea-ice over the Arctic Ocean,
2. increased melting and glacier flow of the Greenland Ice Sheet,
3. "unsticking" of the West Antarctic Ice Sheet from its bed,
4. rapid die-back of Amazon forests,
5. disruption of the Indian monsoon,
6. release of methane, an even more potent greenhouse gas than carbon dioxide, from thawing frozen soils, and
7. a shift to a permanent El Niño–like state.

As with the THC, should any of these occur, the ramifications would be profound: threatening our food production, and causing the extinction and expansion of species and the inundation of coastal cities.

Consider the Greenland and Antarctic ice sheets. The amount of water stored in them is enormous—enough to drown the planet under more than two hundred feet of water. That will not happen anytime soon, but even a tiny reduction in their extent—say, 5 percent—would significantly alter our coastlines. Global sea level is already rising about one-third of a centimeter every year and will rise at least eighteen to sixty centimeters higher in just one long human lifetime, if the rates at which glaciers now flow from land to ocean remain constant. But at least two warming-induced triggers might speed the glaciers up: percolation of lubricating meltwater down to their beds and the disintegration of floating ice shelves that currently pin them onto the continent. If these giants awaken, our best guess is eighty to two hundred centimeters of sea-level rise. That's a lot of water. Most of Miami would either be surrounded by dikes or submerged.

Unfortunately, the presence of sleeping giants makes the steady, predictable growth of anthropogenic greenhouse warming more dangerous, not less. Alarm clocks may be set to go off, but we don't know what their temperature settings are. The science is too new—and besides, we'll never know for sure until it happens. While some economists predicted that rising credit-default swaps and other highly leveraged financial products might eventually bring about an economic collapse, who could have foreseen the exact timing and magnitude of late 2008? As with most threshold phenomena, it is extremely difficult to know just how much poking is needed to disturb the sleep of a giant. Forced to guess, I'd mutter something about decades, or centuries, or never. On the other hand, one of these behemoths may be stirring already: In September 2007, then again in 2008, for the first time in memory nearly 40 percent of the late-summer sea-ice in the Arctic Ocean abruptly disappeared.

Unlike Jack's, the eyes of scientists are slow to adjust to the gloom, but we are beginning to see some outlines and discern not one but many sleeping forms. What is certain is that our inexorable loading of the atmosphere with heat-trapping greenhouse gases increases the likelihood that one or more of them will wake up.

CONSERVING THE CLIMATE: WILL GREENLAND'S MELTING ICE THE DEAL?

———◦———

STEPHEN H. SCHNEIDER

STEPHEN H. SCHNEIDER is a biologist and climatologist at Stanford University and the author of *Laboratory Earth: The Planetary Gamble We Can't Afford to Lose.*

Scientists have been talking about the risks of human-induced climate changes for decades now in places like Congress, scientific conventions, media events, and corporate boardrooms, and at visible cultural extravaganzas like Live Earth. Yet a half-century after serious scientific concerns surfaced, the world is still far from a meaningful deal to implement actions to curb the threats by controlling the offending emissions.

The reason is obvious: Controlling the basic activity—burning fossil fuels—that brought us our prosperity is not going to be embraced by those who benefit from using the atmosphere as a place to dump, for free, their tailpipe and smokestack effluents. Nor will developing economies such as China and India easily give up the techniques we employed to get rich, just because of some threat perceived as distant and not yet certain. To be sure, there is real action at local, state, national, and international levels, but a game-changing global deal is still far from likely. Documented effects, such as loss of the Inuit hunting culture, small island states threatened by inexorable sea-level rise, imminent species extinction in critical places like mountaintops, or a fivefold increase in wildfires in the U.S. West since 1970, have not been game changing—yet. What might change the game?

In order to give up something (the traditional pathway to wealth: burning coal, oil, and gas), nations will have to understand viscerally that they are getting something: protection from unacceptably severe outcomes. The latter has been difficult to achieve, because most scientific assessments are honest, concluding that along with many credible and major risks there are many remaining uncertainties.

We cannot pin down whether sea levels will rise a few feet or a few meters in the next century or two. The former is nasty but relatively manageable with adaptation investments; the latter would mean abandoning coastline installations or cultures where a sizable chunk of humanity lives and works. If we could show scientifically that such a threat was likely, it would be game changing in terms of motivating the kinds of compromises required to achieve the solutions that are currently politically difficult.

This is where the potential for up to seven meters of sea-level rise stored as ice on Greenland will come in to tip us toward meaningful actions. Already, Greenland is apparently melting at an unprecedented rate, way faster than any of our theories or models predicted. But it can be, and has been, argued that this is just a short-term fluctuation, since large changes in ice volume come and go typically on millennial timescales—though mounting evidence from ice cores says there is probably unprecedented melting going on right now. Another decade or two of such scientifically documented acceleration of melting could indeed imply that we will get the unlucky outcome: meters of sea-level rise in the time frame of human infrastructure lifetimes for ports and cities—to say nothing of vulnerable natural places, like coastal wetlands.

Unfortunately, the longer we wait for more confident "proof" of game-changing melt rates in Greenland (or West Antarctica, where another five meters of potential sea-level rise lurks), the higher the risk of passing a tipping point in which the melting becomes an unstoppable, self-driven process. That game change occurrence would force unprecedented retreat from the sea, a major abandonment or rebuilding of coastal civilization, and loss of coastal wetlands. This is a gamble with Laboratory Earth that we can't afford to lose.

CLIMATE WILL CHANGE EVERYTHING

———◇———

WILLIAM CALVIN

WILLIAM CALVIN is affiliate professor emeritus of psychiatry and behavioral sciences at the University of Washington School of Medicine and the author of *Global Fever: How to Treat Climate Change.*

Climate will change our worldview. That each of us will die someday ranks up there, as Marcelo Gleiser points out, with $2 + 2 = 4$ as one of the great certainties of all time. But we are accustomed to think of our *civilization* as perpetual, despite all of the history and prehistory that tells us that societies are fragile. The junior-sized slices of society, such as the church or the corporation, also assumed to outlive the participant, provide us with everyday reminders of bankruptcy. Climate change is starting to provide daily reminders, challenging us to devise ways to build in resiliency, an ability to bounce back when hit hard.

Climate may well force on us a major change in how science is distilled into major findings. There are many examples of the ponderous nature of big organizations and big projects. While I think that the Intergovernmental Panel on Climate Change deserves every bit of its Nobel, the emphasis on "certainty" and the time required for a thousand scientists and a hundred countries to reach unanimous agreement probably added up to a considerable delay in public awareness and political action.

Climate will change our ways of doing science, making some

areas more like medicine with its combination of science and interventional activism, where delay to resolve uncertainties is often not an option. Few scientists are trained to think this way— and certainly not climate scientists, who are having to improvise as the window of interventional opportunity shrinks. Climate will, at times, force a hiatus on doing science as usual, much like what happened during World War II, when many academics laid aside their teaching and research interests to focus intensively on the war effort.

The big working models of fluid dynamics that we use to simulate ocean and atmospheric circulation will themselves be game changing for other fields of dynamics, such as brain processing and decision making. They should be especially important as they are incorporated into economic research. Climate problems will cause economies to stagger, and we have recently seen how fragile they are. Unlike 1997, when currency troubles were forced by a big El Niño and its associated fires in Southeast Asia, recent events show that even without the boat being rocked by external events, our economy can partially crash just from internal instabilities—equivalent to trying to dance in a canoe. Many people will first notice climate change elsewhere via the economic collapse that announces it.

That something as local as a U.S. housing bubble could trigger a worldwide recession shows us just how much work we have to do in "earthquake retrofits" for our economy. Climate-proofing our financial flows will rely heavily on good models of economic dynamics, studies of how things can go badly wrong within a month. With such models, we can test candidates for economic crash barriers.

Finally, climate's challenges will change our perspective on the future. Long-term thinking can be dangerous if it causes us to neglect the short-term hazards. A mid-century plan for emissions reduction will be worthless if the Amazon rain forest burns down during the next El Niño.

MOLECULAR MANUFACTURING AND CLIMATE CHANGE

———— ◦ ————

ERIC DREXLER

ERIC DREXLER is an engineer and nanotechnologist, and the author of *Engines of Creation: The Coming Era of Nanotechnology*.

I see great change flowing from the spread of knowledge of two scientific facts, one simple and obvious, the other complex and tangled in myth. Both are crucial to understanding the climate-change problem and what we can do about it.

First, the simple scientific fact: Carbon stays in the atmosphere for a long time.

To many readers, this is nothing new, yet most who know this make an elementary mistake. They think of carbon as if it were sulfur, with pollution levels that rise and fall with the rate of emission: Cap sulfur emissions and pollution levels stabilize; cut emissions in half, cut the problem in half. But carbon is different. It stays aloft for about a century. It accumulates. Cap the rate of emissions and the levels keep rising; cut emissions in half and levels will still keep rising. Even deep cuts won't reduce the problem, only the problem's rate of growth.

In the bland language of the Intergovernmental Panel on Climate Change, "only in the case of essentially complete elimination of emissions can the atmospheric concentration of CO_2 ultimately be stabilized at a constant [far higher!] level." This heroic feat would require new technologies and the replacement of today's installed infrastructure for power generation, transportation, and manufacturing. This seems impossible. In the real world Asia is industrializing, most new power plants burn coal, and emissions are accelerating, increasing the rate of increase of the problem.

The second fact (complex and tangled in myth) is that this seemingly impossible problem has a correctable cause: The human race is bad at making things, but physics tells us that we can do much better.

This will require new methods for manufacturing, methods that work with the molecular building blocks of the stuff that makes up our world. In outline (says physics-based analysis), nanoscale factory machinery operating on well-understood principles could be used to convert simple chemical compounds into beyond-state-of-the-art products and do this quickly, cleanly, inexpensively, and with a modest energy cost. If we were better at making things, we could make those machines, and with them we could make the products that would replace the infrastructure that is causing the accelerating and seemingly irreversible problem of climate change.

What sorts of products? Returning to power generation, transportation, and manufacturing: Picture roads resurfaced with solar cells (a tough, black film), cars that run on recyclable fuel (sleek, light, and efficient), and car factories that fit in a garage. We could make these easily, in quantity, if we were good at making things.

Developing the required molecular manufacturing capabilities will require hard but rewarding work on a global scale, converting scientific knowledge into engineering practice to make tools we can use to make better tools. The aim that physics suggests is a factory technology with machines that assemble large products from parts made of smaller parts (made of smaller parts, and so on) with molecules as the smallest parts and the smallest machines only a hundred times larger.

The basic science to support this undertaking is flourishing, but the engineering has gotten a slow start, and for a peculiar reason: The idea of using tiny machines to make things has been burdened by an overgrowth of mythology. According to fiction and pop culture, it seems that all tiny machines are robots made of diamond, and they're *dangerous magic*—smart and able to do almost anything for us, but apt to swarm and multiply and maybe eat everything, probably including your socks.

In the real world, manufacturing does indeed use robots, but these are immobile machines that work in an assembly line, putting part A into slot B, again and again. They don't eat, reproduce, or go out on strike, and making them smaller wouldn't make them any smarter.

There is a mythology in science, too, but of a more sober sort—not a belief in glittery nanobugs but a skepticism rooted in mundane misconceptions about whether nanoscale friction and thermal motion will sabotage nanomachines, and whether practical steps can be taken in laboratories today. (No, and yes.) This mythology, by the way, seems regional and generational; I haven't encountered it in Japan, India, Korea, or China, and it is rare among the rising generation of researchers in the United States.

The U.S. National Academies have issued a report on molecular manufacturing, and it calls for funding experimental research. Battelle, with several U.S. National Laboratories, has studied paths forward and prepared a roadmap that suggests research directions. This knowledge will spread, and will change the game.

I should add one more fact about molecular manufacturing and the climate-change problem: If we were good at making things, we could make efficient devices able to collect, compress, and store carbon dioxide from the atmosphere, and we could make solar arrays large enough to generate enough power to do this on a scale that matters. A solar array area that, if aggregated, would fit in a corner of Texas could generate three terawatts. In the course of ten years, three terawatts would provide enough energy to remove all the excess carbon the human race has added to the atmosphere since the Industrial Revolution began. As far as carbon emissions are concerned, this would fix the problem.

THE MASTERY OF CLIMATE

――◆◇◆――

STEWART BRAND

STEWART BRAND is the founder of *The Whole Earth Catalog*, cofounder of The Well and the Global Business Network, and the author of *The Clock of the Long Now: Time and Responsibility: The Ideas Behind the World's Slowest Computer*.

To take mastery of climate as we once took mastery of fire, then of genetics (agriculture), then of communication (music, writing, math, maps, images, printing, radio, computers), will require mathematics we don't have, physics and biology we don't have, and governance we don't have.

Our climate models, sophisticated and muscular as they are (employing more teraflops than any other calculation), still are just jumped-up weather prediction models. The real climate system has more levels and modes of hyperconnected nonlinearity than we can yet comprehend or ask computers to replicate, because so far we lack the math to represent climate dynamics with the requisite variety to control it. Acquiring that math will change everything.

Materials scientist and engineer Saul Griffith estimates that humanity must produce thirteen terawatts of greenhouse-free energy in order to moderate global warming to a just tolerable increase of 2°C. (Civilization currently runs on about sixteen terawatts of energy, most of it from burning fossil fuels.) Griffith calculates that deploying current clean technologies—nuclear, wind, geothermal, biofuels, and solar technology—to generate thirteen terawatts would cover an area the size of Australia. It is imaginable but not feasible. Just improving the engineering of nuclear and solar won't get us what we need; new science is required. The same goes for biofuels: The current state of genetic engineering is too crude to craft truly efficient organisms for sequestering carbon and generating usable energy. The science of molecular biology has to advance by leaps. Applied science that powerful will change everything.

Climate change is a global problem that cannot be fixed with global economics, which we have; it requires global governance, which we don't have. Whole new modes of international discourse, agreement, and enforcement must be devised. How are responsibilities for legions of climate refugees to be shared? Who decides which geoengineering projects can go forward? Who pays for them? Who adjudicates compensation for those harmed? How are free riders dealt with? Humans have managed commons before—fisheries, irrigation systems, fire regimes—but never on this scale. Global governance will change everything.

Of course, these radical adjustments may not happen, or not happen in time, and then climate will shift to either a chaotic mode or a different stable state, with the carrying capacity for just a fraction of present humanity, and that will really change everything.

THE USE OF NUCLEAR WEAPONS AGAINST A CIVILIAN POPULATION

———◁◦▷———

LAWRENCE KRAUSS

LAWRENCE KRAUSS is a physicist at Arizona State University and the author of *Hiding in the Mirror: The Quest for Alternate Realities.*

"The release of atom power has changed everything except our way of thinking." So said Albert Einstein sixty-four years ago, following the Hiroshima and Nagasaki bombings at the end of World War II. Forced to choose a single game changer, I have turned away from the fascinating scientific developments I might like to see and will instead focus on the one game changer I hope I will never witness but nevertheless expect will occur during my lifetime: the use of nuclear weapons against a civilian population.

Whether used by one government against the population of another or by a terrorist group, the detonation of even a small nuclear explosive—similar in size, for example, to the one that destroyed Hiroshima—would alter the economies, politics, and lifestyles of the first world in a way that would make the effects of the 9/11 attacks seem trivial. I believe that the possible use of nuclear weapons remains one of the biggest dangers of this century. It is remarkable that more than sixty years have passed without their use, but the clock is ticking. I fear that Einstein's admonition is just as true today as it was then and that we are unlikely to go another half-century with impunity—at least without confronting the need for a global program of disarmament that goes far beyond the current Nuclear Non-Proliferation and strategic arms treaties.

Following forty years of Mutual Assured Destruction, with the two superpowers like two scorpions in a bottle, each held at bay by the certainty of the destruction that would occur at the first whiff of nuclear aggression on the part of the other, we have become complacent. Two generations have come to maturity in a world where nuclear weapons have not been used. The Nuclear Non-Proliferation Treaty has been largely ignored, and not just by nascent nuclear states like North Korea or India or Pakistan or pre-nuclear wannabes like Iran. Together, the United States and Russia possess twenty-six thousand of the world's twenty-seven thousand known nuclear warheads—this in spite of the NNPT's strict requirement for these countries to significantly reduce their arsenals. Each country has perhaps a thousand warheads on hair trigger full alert—even though there is currently no strategic utility associated with such a posture.

Ultimately, what concerned Einstein, and is of equal concern today, is that first use of nuclear weapons cannot be justified on either moral or strategic grounds. Nevertheless, it may surprise some people to learn that the United States has no strict anti-first-use policy. In fact, in its 2002 Nuclear Posture Review, the United States declared that nuclear weapons "provide credible military options to deter a wide range of threats," including "surprising military developments."

While we spend $10 billion a year on flawed ballistic missile defense systems against nonexistent threats, the slow effort to disarm means that thousands of nuclear weapons remain in regions that are unstable and could, in principle, be accessed by well-organized and well-financed terrorist groups. We have not used a noticeable fraction of the money spent supposedly defending ourselves against ballistic missiles to instead outfit ports and airports to detect nuclear devices that might be smuggled into this country in containers.

Will it take a nuclear attack on a civilian population to stir a change in thinking? The havoc wreaked on what we now call the civilized world, no matter where a nuclear confrontation takes place, would be orders of magnitude greater than anything we have experienced since World War II. Moreover, as recent calculations have demonstrated, even a limited nuclear exchange—say, between India and Pakistan—could have a significant global effects on world climates and growing seasons for almost a decade.

I sincerely hope that whatever initiates a global realization that the existence of large nuclear stockpiles is a threat to everyone on the planet, changing the current blind, business-as-usual mentality permeating global strategic planning, does not result from a nuclear tragedy. But physics has taught me that the world is the way it is whether we like it or not. And my gut tells me that to continue to ignore the likelihood that a game changer exceeding our worst nightmares will occur in this century is one way to encourage that possibility.

DEPLOYMENT OF A SIGNIFICANT ROGUE NUCLEAR DEVICE

————◆◇◆————

GERALD HOLTON

GERALD HOLTON is Mallinckrodt Professor of Physics, professor emeritus of the history of science at Harvard University, and coeditor, with Peter L. Galison and Silvan S. Schweber, of *Einstein for the 21st Century: His Legacy in Science, Art, and Modern Culture.*

An answer can be given in one sentence: the intentional, hostile deployment, whether by a state, a terrorist group, or other individuals, of a significant nuclear device.

ACCIDENTAL NUCLEAR WAR

◄◇►

MAX TEGMARK

MAX TEGMARK is a physicist and cosmologist at MIT and scientific director of the Foundational Questions Institute.

A serial killer is on the loose! A suicide bomber! Beware the West Nile virus!

Headline-grabbing scares are better at generating fear, but boring old cancer is more likely to do you in. Although you have less than a 1 percent chance per year to get it, live long enough and it has a good chance of dispatching you in the end. As does accidental nuclear war.

During the half-century we have been tooled up for nuclear Armageddon, there has been a steady stream of false alarms that could have triggered all-out war, with causes including computer malfunction, power failure, faulty intelligence, navigation error, bomber crash, and satellite explosion. Gradual declassification of records has revealed that some of those occurrences carried greater risk than was appreciated at the time. For example, it became clear only in 2002 that during the 1962 Cuban missile crisis, the USS *Beale* depth-charged an unidentified submarine that was in fact Soviet, and whose officers argued over whether to retaliate with a nuclear torpedo.

Despite the end of the cold war, the risk of nuclear attack has arguably grown in recent years. Inaccurate but powerful ICBMs undergirded the stability of Mutual Assured Destruction, because a first strike could not prevent massive retaliation. The shift toward more accurate missile navigation, shorter flight times, and better enemy submarine tracking erodes this stability. A successful missile defense system would complete the erosion process. Both Russia and the United States retain their "launch-on-warning" strategy, requiring launch decisions to be made on five- to fifteen-minute timescales, when complete information may be unavailable. On January 25, 1995, Russian president Boris Yeltsin came within minutes of initiating a full nuclear strike on the United States because of an unidentified Norwegian scientific rocket. Concern has been raised over a recent U.S. project to replace the nuclear warheads on two of the twenty-four D5 ICBMs carried by Trident submarines with conventional warheads for possible use against Iran or North Korea. Russian early-warning systems would be unable to distinguish them from nuclear missiles, expanding the possibilities for unfortunate misunderstanding. Other worrisome scenarios include deliberate malfeasance by military commanders prompted by mental instability and/or fringe political or religious agendas.

But why worry? Surely, if push comes to shove, won't reasonable people step in and do the right thing, just as they have in the past?

Nuclear nations do indeed have elaborate countermeasures in place, just as our bodies do against cancer. Our bodies can normally deal with isolated harmful mutations, and it appears that fluke coincidences of as many as four mutations may be required to trigger certain cancers. Yet if we roll the dice enough times, shit happens; Stanley Kubrick's dark nuclear comedy *Dr. Strangelove* illustrates this with a triple coincidence.

Accidental nuclear war between two superpowers may or may not happen in my lifetime, but if it does, it will obviously change everything. The alterations in climate we are now discussing pale in comparison with nuclear winter, and the current economic turmoil is nothing compared with the resulting global crop failures, infrastructure collapse, and mass starvation, with survivors succumbing to hungry armed gangs systematically pillaging from house to house. Do I expect to see this in my lifetime? I'd give it about 30 percent, putting it roughly on par with me getting cancer. Yet we devote much less attention and many fewer resources to reducing this risk than we do to lowering our cancer risk.

THE BREAKDOWN OF ALL COMPUTERS

ANTON ZEILINGER

ANTON ZEILINGER is a professor of physics at the University of Vienna and scientific director of the Institute of Quantum Optics and Quantum Information, Austrian Academy of Sciences.

Someday all semiconductors will break down, and therefore all computers, as (with the exception of historical instruments) no computers exist today that are not based on semiconductor technology. The breakdown will be caused by a giant electromagnetic pulse (EMP) created by a nuclear explosion outside Earth's atmosphere. It will cover large areas on Earth, up to the size of a continent. Where it will happen is unpredictable. But it will happen, since it is extremely unlikely that we will be able to get rid of all nuclear weapons, and the probability for it to happen at any given time will never be zero.

The implications of such an event will be enormous. If it happens to one of our technology-based societies, literally everything will break down. You will realize that none of your phones work. There is no way to find out via the Internet what happened. Your car will not start, as it is also controlled by computer chips—unless you are lucky enough to own an antique car. Your local supermarket will be unable to get new supplies. There will be no trucks operating, no trains, no electricity, no water supplies. Society will completely break down.

There will be small exceptions in those countries where military equipment has been hardened against EMPs, making the army available for emergency relief. In some countries, even some emergency civilian infrastructure has been hardened against EMPs. But these are exceptions. Most governments simply ignore the danger.

THE GROWING PERCEPTION OF A CLASH BETWEEN SAFETY AND LIBERTY

DAN SPERBER

DAN SPERBER is a social and cognitive scientist, director of research at the Centre National de la Recherche Scientifique, Paris, and the author of *Explaining Culture: A Naturalistic Approach*.

From the Neolithic Revolution to the Information Age, the major changes in the human condition—none of them changing everything, needless to say—have been consequences of new technologies. There is now a glut of new technologies in the offing that will alter the way we live more rapidly and radically than anything before, in ways we cannot properly foresee. I wish I could just wax lyrical about some of the developments we can at least sensibly speculate about, but others will do so more competently. So let me focus on the painfully obvious that we would rather not think about.

Many new technologies can provide new weapons or new ways to use old ones. Access to these technologies is easier every day. We should expect with virtual certainty that in the near future, atomic, chemical, and biological weapons of mass destruction will be used in a variety of conflicts. The most important change this will bring about is not that so many will die. Hundreds of thousands have died all these years, in wars and natural catastrophes, with an unspeakable impact on the population affected. But alas, massacres and other forms of collective death have been part and parcel of the human condition. This time, however, many of the victims will belong to powerful modern societies that, since World War II, have on the whole been spared. People in these societies are, neither less nor more than the usual poorer and powerless victims of massive violence, entitled to live full decent lives and have a right to fight for this. What may bring about radical changes is that they will be in a much stronger position to exert and possibly abuse this right. Recent large-scale murderous attacks have resulted in the acceptance of fewer limits on executive power, the curtailing of civil rights, the waging of preventive warfare, and ethnically targeted public suspicion. In the future, people who will have witnessed even more dire events at close quarters may well support even more drastic measures. I am not discussing here the rationality of fears to come or the extent to which they are likely to be biased and manipulated. I just assume that, for good or bad reasons, they will weigh in in favor of limitations to the liberties of individuals and to the independence of countries.

One must hope that—in part, thanks to the changes brought about by novel technologies—new forms of social and political understanding and action will develop to help address the root issues that otherwise might give rise to ever-more lethal conflicts. Still, while more and more powerful technologies are becoming more and more accessible, there is no reason to believe that humans are becoming commensurately wiser and more respectful of one another's rights. There will be, at least in most people's perception, a direct clash between their safety and their liberty, and even more between the safety and the liberty of others. The history of this century—our history, that of our children and grandchildren—will in good part be that of the ways in which this clash is played out, or overcome.

ADOPTING RATIONALITY AND SUSTAINABILITY

—◇—

PATRICK BATESON

PATRICK BATESON is professor emeritus of ethology at the University of Cambridge and the author of *Design for a Life: How Biology and Psychology Shape Human Behavior.*

"The release of atom power has changed everything except our way of thinking." So said Einstein. Whether or not he foresaw the total destruction of the world, the thought gave rise to a joke (albeit a sick one) that humans would never make contact with civilizations in other parts of the universe. Either those civilizations were not advanced enough to decode our signals or they were more advanced than us, had developed nuclear weapons, and had destroyed themselves. The chances would be vanishingly small that their brief window of time between technological competence and oblivion would coincide with ours.

I never understood the policy of deterrence that justified the nuclear arms race. The coherence of such a view depended utterly on the maintenance of human rationality. Suppose that people whose concern for personal safety or the welfare of others is subordinated to religious or ideological belief rule a country in possession of nuclear weapons. The whole notion of deterrence collapses.

I usually regard myself as an optimist: Tomorrow will be better than today. My naïve confidence has been dented by advancing age and by the growing number of reality checks that point to trouble ahead. Even if the red mists of anger or insanity do not unleash the total destruction of our way of life, the prognosis for the survival of human civilization is not good. However much we believe in technical fixes that will overcome the problems of diminishing resources, the planet is likely to be overwhelmed by the sheer number of people who inhabit it and by the conceit that economic growth for everybody is the only route to well-being. The uncontrolled greed of the developed world has taken a sharp knock in the recent credit crunch, but how do you persuade affluent people to accept an overall reduction in their standard of living? Which government of any stripe is going to risk its future by enforcing the unpopular policies that are already needed? On this front, the prognosis might not be too bad, since crises do bring about change.

The Yom Kippur War of 1973 led to a dramatic reduction in the oil supply; in the UK, gasoline rationing was swiftly introduced and everybody was required to save fuel by driving no faster than fifty miles per hour. The restraint disappeared, of course, as soon as oil started to flow again, but the experience showed that people will uncomplainingly change their behavior when they are required to do so and understand the justification.

Growth of the human population must be one of the major threats to sustainability of resources such as drinking water and food. Here again, the prognosis does not have to be wholly bad if farsighted wisdom prevails. If the gross domestic product of each country of the world is plotted on graph paper against average family size in that country, the correlation is almost perfectly negative. (The outliers provided by rich countries with large average family sizes are almost exclusively those places in which women are treated badly.) The evidence suggests that if we wish to reduce population growth, every effort should be made to boost the GDP of the poorest countries of the world. This is an example of how economic growth in some countries and overall benefit to the world can proceed hand in hand, but the richest countries will have to pay the price.

Another, darker thought is that the human population might be curbed by its own stupidity and cupidity. I am not now thinking of conflict but of the way in which endocrine disruptors are poured unchecked into the environment. Suddenly males might be feminized by the countless number of artificial products that simulate the action of female hormones—sufficiently so that reproduction becomes impossible. For some, the irony would be delicious: the ultimate feedback mechanism, unforeseen by Malthus, that places a limit on population growth.

Sustainability requires that we pass on to the next generation the resources (or some equivalent) that we received from our forebears. It may be a pipe dream, given the way we think.

FUSION EXPECTATIONS

———◆———

ROGER HIGHFIELD

ROGER HIGHFIELD is the editor of *New Scientist* and the coauthor, with Ian Wilmut, of *After Dolly: The Promise and Perils of Cloning*.

Now, this idea will end the energy crisis and curb climate change at a stroke. I am confident in what I say, because a lot of clever people have said it again and again—and again—for more than half a century. Since the heady, optimistic days when scientists first dreamed of taming the power of the sun, fusion energy has remained tantalizingly out of reach.

It will take us between twenty and fifty years to build a fusion power plant. That is what glinty-eyed scientists announced at the height of the cold war. Their modern equivalents are still saying it. And I am going to say it once again, because it really could—and will—make a difference.

Fusion power could be a source of energy that would have a greater effect on humankind than landing the first man on the moon. The reason is, as one former UK government chief scientist liked to put it, that the lithium from one laptop battery and the deuterium from the water in your bath would generate enough energy to power a single citizen for thirty years. And overall, fusion reactors would create fewer radioactive waste problems than their fission sisters.

The skeptics have long sneered that the proponents of fusion power are out of touch with reality. As the old joke goes, fusion is the power of the future, and it always will be. But this is one energy bet that must pay off, given the failure of the Kyoto Protocol.

There are good reasons to be hopeful. In Cadarache, France, construction is under way of ITER, the International Thermonuclear Experimental Reactor. *(Iter* means "the way" in Latin; cynics carp that it can also mean "journey" and a bloody long one, too.) This project will mark a milestone in fusion development; other solid bets are being placed, notably using high-power lasers to kickstart the fusion process.

Greens will complain that the money would be better spent on renewables, but if this unfashionable gamble pays off, the entire planet will be the winner. Imagine the patent squabbles when engineers finally figure out how to make fusion economic. Think of the seismic implications for energy research and alleviating poverty in the developing world. Consider the enormous implications for holding back climate change. We are about to catch up with the receding horizon of fusion expectations.

GREEN OIL

◄◦►

ALUN ANDERSON

ALUN ANDERSON is a senior consultant and former editor and publishing director of *New Scientist*.

Green oil is the development that will utterly change the world, and it will arrive in the next few decades. Oil that we take out of the ground and burn is going to be replaced by oil that we grow.

Biofuels based on corn are our first effort to grow green oil, but they have clearly not succeeded. Current biofuels take too much land and too much energy to grow, and too much of that energy goes into building parts of a plant that we can't easily convert into fuel. The answer will come from simple engineered organisms that can soak up energy in a vat in any sunny spot and turn that sunlight straight into a precursor for fuel, preferably a precursor that can go straight into an existing oil refinery.

The impacts of such a development are staggering. The power balance of the world will be completely changed. Petro-dictatorships, where an endless flow of oil money keeps the population quiet, will no longer be able to look forward to oil at $50, $100, $150, and so on a barrel as oil supplies tighten. Power will be back in the hands of innovators rather than resource owners. The quest for dirty oil in remote and sensitive parts of the world, whether the Arctic or the Alberta tar sands, will not make economic sense, and the environment will gain. The burning of gasoline in automobiles will no longer add much to the amount of carbon dioxide in the atmosphere, as the fuel will have soaked up an almost equal amount of carbon dioxide while it was being grown. The existing networks for delivering fuel to transportation (the hundred thousand gas stations in the Unites States, for example) won't become redundant (as they would if we switched to electric autos), making plans for cutting emissions much less difficult.

Will the green oil come from algae, bacteria, archaea, or something else? I don't know. Oil is a natural product arising from the transformation of plant material created by the capture of light. As it is a transformation in nature, we can replicate it— not necessarily directly, but to arrive at a similar result. It is not magic.

Scientists around the world have seen the prize, and hundreds of millions are going into start-up companies. There is a nice twist to this line of investment. Despite the ups and downs, the long-term trend for the price of oil is up. That means the size of the prize for replacing oil is going up, while the size of the challenge is going down. Replacing $20-a-barrel oil would be difficult, but replacing $100-a-barrel oil is much easier.

There is an old saying: "The Stone Age didn't end because we ran out of stone. Someone came up with a better idea." The better idea is coming.

ATTEMPTS AT GEOENGINEERING

—◁◦▷—

OLIVER MORTON

OLIVER MORTON is chief news and features editor of *Nature* and the author of *Eating the Sun: How Plants Power the Planet*.

It is quite likely that we will at some point see people starting to make deliberate changes in the way the climate system works. When they do, they will change the world—though not necessarily (or only) in the way they intended to.

Geoengineering technologies for counteracting some aspects of anthropogenic climate change—such as putting long-lived aerosols into the stratosphere as volcanoes do, or changing the lifetimes and reflective properties of clouds—have to date been shunned by the majority of climate scientists, largely because of the moral hazard involved: Any sense that the risks of global warming can be taken care of by such technology weakens the case for reducing carbon dioxide emissions.

I expect to see this unwillingness recede dramatically in the next few years, and not only because of the post–Lehman Brothers bashing given to the idea that moral hazard is something to avoid at all costs. As people come to realize how little has actually been achieved so far on the emissions-reduction front, quite a few are going to freak out. Some of those who freak will have money to spend, and with money and the participation of a larger cadre of researchers, the science and engineering required for the serious assessment of various geoengineering schemes might be developed fairly quickly.

Why do I think those attempts will change the world? Geoengineering is not, after all, a panacea. It cannot precisely cancel out the effects of greenhouse gases, and it is likely to have knock-on effects on the hydrological cycle that may well be unwelcome. Even if the benefits outweigh the costs, the best-case outcome is unlikely to be more than a grace period in which the most excessive temperature changes are held at bay. Reducing carbon dioxide emissions will continue to be necessary, in part because of the problem of ocean acidification and in part because a lower-carbon-dioxide climate is vastly preferable to one that stays teetering on the brink of disaster for centuries, requiring constant tinkering to avoid slipping into greenhouse hellishness.

So geoengineering will not "solve" climate change. Nor will it be an unprecedented human intervention in the Earth system. It will be a huge undertaking, but hardly more momentous in absolute terms than our replacement of natural ecosystems with farmed ones, or our commandeering of the nitrogen cycle, or the wholesale havoc we have wreaked on marine food webs, or the amplification of the greenhouse effect itself.

What I see as world changing about this technology is not the extent to which it changes the world, but that it does so on purpose. To live in a world subject to purposeful, planetwide change will not, I think, be quite the same as living in one being messed up by accident. Unless geoengineering fails catastrophically (which would be a pretty dramatic change in itself), the relationship between people and their environment will have changed profoundly. The line separating the natural from the artificial is itself an artifice and one that changes with time. But this change, different in scale and not necessarily reversible, might finish off the idea of the natural as a place or time or condition that could ever be returned to. This would not be "the end of nature," but it would be the end of a view of nature that has great power and without which some would feel bereft. The clouds and the colors of the noontime sky and of the setting sun will feel different if they have become, to some extent, a matter of choice.

And that choice is itself another aspect of the great change: Who chooses, and how? All climate change, whether intentional or not, has different outcomes for different regions, and geoengineering is in many ways just another form of climate change. So for some it will likely make the situation worse. If it does, does that constitute an act of war? An economic offense for which others will insist on reparations? Just one of those things that the stronger do to the weaker?

Critics of geoengineering approaches are right to stress this governance problem. Where they tend to go wrong is in ignoring the fact that we already have a climate governance problem: The mechanisms currently in place to "avoid dangerous climate change," as the United Nations' Framework Convention on Climate Change puts it, so far have not delivered the goods. A system conceived with geoengineering in mind would need to be one that held countries to the consequences of their actions in new ways, and that might strengthen and broaden approaches to emissions reduction, too. But there will always be an asymmetry, and it is an important one: To do something about emissions, a significant number of large economies will have to act in concert. Geoengineering can be unilateral. Any medium-sized nation could try it.

In this, as in other ways, geoengineering issues look oddly like nuclear issues. There, too, a technological stance by a single nation can have global consequences. There, too, technology has reset the boundaries of the natural in ways that can provoke visceral opposition. There, too, there are a discourse of transcendence and a tendency to hubris that need to be held in check. And there, too, the technology has brought with it dreams of new forms of governance. In the light of Trinity, Hiroshima, and Nagasaki, many saw some sort of world government as a moral imperative, a historical necessity, or both. It turned out not to be, and the control of nuclear weapons and ambitions has remained an ad hoc thing—a mixture of treaties, deterrence, various suasions, and occasional direct action—that is unsatisfactory in many ways, though not as yet a complete failure. A geoengineered world may end up governed in a similarly piecemeal way—and bring with it a similar open-ended risk of destabilization and even disaster.

The world has inertia and complexity. It changes, and it can be changed—not always quickly, and not necessarily controllably, and not all at once. But within those constraints, geoengineering will bring changes, and it will do so intentionally. And that intentional change in the relationship between people and planet might be the biggest change of all.

WHY DON'T RUNNING SHOES BIODEGRADE?

DANIEL GOLEMAN

DANIEL GOLEMAN is a psychologist and the author of *Ecological Intelligence: How Knowing the Hidden Impacts of What We Buy Can Change Everything.*

Every man-made object—all the things in our homes and workplaces—has an invisible backstory, a litany of sorry impacts over the course of the journey from manufacture to use to disposal. Take running shoes.

Despite the bells and whistles meant to make one brand of running shoe appeal more than another, at base they all reduce to three parts. The shoe's upper consists of nylon with decorative bits of plastics or synthetic leather; the "rubber" outsole for most shoes is a petroleum-based synthetic; so is the spongy midsole, composed of ethylene vinyl acetate. Like any petrochemical widget, manufacturing the soles produces unfortunate by-products, among them benzene, toluene, ethyl benzene, and xylene. In environmental health circles, these are known as the Big Four toxics, being variously carcinogens, central nervous system disrupters, and respiratory irritants, among other biological shortcomings.

Those bouncy air pockets in some shoe soles contain an ozone-depleting gas. The decorative bits of plastic piping harbor polyvinyl chloride (PVC), which endangers the health of workers who make it and contaminates the ecosystems around the dumps where we eventually send our shoes. The solvents in glues that bind outsole to midsole can damage the lungs of the workers who apply it. Tanning leather for shoe tops can expose workers to hexavalent chromium and other carcinogens.

I remember my high school chemistry teacher's enthusiasm for the chemical reaction that rendered nitrogen fertilizer from ammonia (he moonlighted in a local fertilizer factory); we never heard a word about eutrophication, the dying of aquatic life due to fertilizer runoff that creates a frenzy of algae growth, depleting the water's oxygen. Likewise, coal-burning electric plants seemed a marvel when first deployed: cheap electricity from a virtually inexhaustible source. Who knew about respiratory disease from particulates, let alone global warming?

The full list of adverse impacts on the environment or on the health of those who make or use any product can run into hundreds of such details. The reason: Almost all of the manufacturing methods and industrial chemicals in common use today were invented in a day when little or no attention was paid to their negative impacts on the planet or its people.

We have inherited an industrial legacy from the twentieth century that no longer meets the needs of the twenty-first. As we awaken from our collective naïveté about such hidden costs, we are reaching a pivot point where we can question hidden assumptions. We can ask, for example, "Why not have running shoes that are not just devoid of toxins, but also can eventually be tossed out in a compost pile to biodegrade?" We can rethink everything we make, developing alternative ingredients and processes with far less—or ideally, no—adverse health or environmental impacts.

The singular force that can drive this transformation of every man-made thing for the better is neither government fiat nor the standard tactics of environmentalists, but radical transparency in the marketplace. If we, as buyers, can know the actual ecological effects of the stuff we buy at the point of purchase and can compare those effects with those of competing products, we can make better choices. The means for such radical transparency has already been launched. Software innovations now allow any of us to access a vast database about the hidden harms in whatever we are about to buy, and to do this where it matters most: at the point of purchase. As we stand in the aisle of a store, we can know which brand has the fewest chemicals of concern or the better carbon footprint. In the beta version of such software, you click your cell phone's camera on a product's bar code and get an instant readout of how this brand compares with competitors on any of hundreds of environmental, health, or social consequences. In a planned software upgrade, that same comparison would go on automatically with whatever you buy on your credit card, and suggestions for better purchases next time you shop would routinely come your way by e-mail.

Such transparency software converts shopping into a vote, letting us target manufacturing processes and product ingredients we want to avoid, and rewarding smarter alternatives. As enough of us apply these decision rules, market share will shift, giving companies powerful, direct data on what shoppers want—and want to avoid.

Creating a market force that continually leverages ongoing upgrades throughout the supply chain could open the door to immense business opportunities over the next several decades. We need to reinvent industry, starting with the most basic platforms in industrial chemistry and manufacturing design. And that would change everything.

THE SHIFT FROM HARVESTING TO MANUFACTURING ENERGY

ANDRIAN KREYE

ANDRIAN KREYE is Arts and Ideas Editor, *Süddeutsche Zeitung*, Munich.

It should be an easy transition. Instead of thinking about energy as a commodity to harvest, new sources of power will be manufactured. The medieval quest for new sources of that life force called energy will be over, including all those white knights on horses conquering the wild lands where those sources happen to be. Technologically this will mean a shift from an energy industry dominated by geologists and engineers to a wave of innovations driven by biologists and chemists.

The thought process itself has already been set in motion. The surge of first-generation biofuels has been based on the idea of renewable sources of energy. Still, most alternative energies, such as solar power and wind power, remain based on the old way of thinking—about harvesting. Most biofuels are preceded by a literal harvest of crops. Craig Venter's work on a microorganism that can transform carbon dioxide, sunlight, and water into fuel is already jumping quite a few steps ahead.

This new approach will drastically reduce the EIOER (energy input versus energy return) formula that has so far slowed down the commercial viability of most innovations in the search for alternative sources of energy. Any fuel that can be synthetically "grown" in a lab or factory will be economically much more viable for mass production than the conversion of sunlight, wind, or agricultural goods.

Laboratory-based production of synthetic sources of energy will also end the geopolitical dependencies now tied to the consumption of power and thus change the course of recent history in the most dramatic fashion. It will eliminate the sources of many current and future conflicts, first and foremost in the Gulf region, but also in the northern part of South America, in the Black Sea region, and in the increasingly exploitable Arctic.

The introduction of biological processes into the energy cycle will also minimize the impact of energy consumption on the environment. If made available cheaply, possibly as an open-source endeavor, it will allow emerging nations to develop new arable land and create wealth while avoiding conflict and environmental negligence.

There could, of course, be downsides to the emergence of new sources of energy. Transitions are never easy, no matter how benign or progressive. The loss of economical and political power by oil- and gas-producing nations and corporations could become a new, if temporary, source of conflict. Unforeseen dangers in the production might emerge that will adversely affect environmental and public health. New monopolies could be formed.

The shift from harvesting to manufacturing energy would not only affect the economy, geopolitics, and the environment: Turning humankind from mere harvesters of energy into manufacturers of energy would lead to a whole new way of thinking, one that in turn could lead to even greater innovations, because every form of economic and technological empowerment always initiates leaps that go way beyond the practical application of new technologies. It's hard to predict where a new mindset will lead. One thing is for sure: It almost always leads to new freedom and enlightenment.

THE ANTHROPOSPHERE

---◄◊►---

NICHOLAS A. CHRISTAKIS

NICHOLAS A. CHRISTAKIS is an internist and social scientist at Harvard University.

We will create life from inanimate compounds, and we will find life on Mars or in space. But the life that more immediately interests me lies between these extremes, in the middle range we all inhabit between our genes and our stars. It is the thin bleeding line within the thin blue line, the anthroposphere within the biosphere, the part of the material world in which we live out our lives. It is us.

And we are rapidly and inexorably changing. I do not mean that our numbers are exploding, a topic that has been attracting attention since Malthus. I mean a very modern and massive set of changes in the composition of the human population.

The global population stood at one million in 10,000 BC, fifty million at 1000 BC, and three-hundred-ten million in AD 1000. It stood at about one billion in 1800, 1.65 billion in 1900, and six billion in 2000. Analysis of these macrohistorical trends in human population usually focuses on this population growth and on the "demographic transition" underlying it.

During the first stage of the demographic transition, life—as Hobbes rightly suggested—was nasty, brutish, and short. There was a balance between birthrates and death rates, and both were very high (thirty to fifty per thousand people per year). The human population grew less than 0.05 percent annually, with a doubling time of over a thousand years. This state of affairs was true of all human populations everywhere, until the late eighteenth century.

Then, during the second stage, the death rate began to decline—first in northwestern Europe, but then spreading over the next hundred years to the south and east. The decline in the death rate was due initially to improvements in food supply and in public health, both of which reduced mortality, particularly in childhood. As a consequence, there was a population explosion.

During the third stage, birthrates dropped for the first time in human history. The prior decline in childhood mortality probably prompted parents to realize that they did not need as many children; and increasing urbanization, increasing female literacy, and (eventually) contraceptive technology also played a part.

Finally, during the fourth stage—in which the developed world currently finds itself—there is renewed stability. Birth and death rates are again in balance, but now both are relatively low. Causes of mortality have shifted from the premodern pattern dominated by infectious diseases, perinatal diseases, and nutritional diseases to one dominated by chronic diseases, mental illnesses, and behavioral conditions.

This broad story, however, conceals as much as it reveals. There are other demographic developments worldwide beyond the increasing overall size of the population, developments that are still unfolding and that matter much more. Changes in four aspects of population structure are key: (1) sex ratio, (2) age structure, (3) kinship systems, and (4) income distribution.

Sex ratios are becoming increasingly unbalanced in many parts of the world, especially in China and India (which account for 37 percent of the global population). The normal sex ratio at birth is roughly one-hundred-six males for every hundred females, but it may soon be as high as one-hundred-twenty for young people in China, or as high as one-hundred-eleven in India. This shift, much discussed, may arise from preferential abortion or the neglect of baby girls relative to boys. Gender imbalance may also have other determinants, such as large-scale migration of one or the other sex in search of work. This shift has numerous implications. For example, given the historical role of females as caregivers to elderly parents, a shortage of women to fill this role will induce large-scale social adjustments. Moreover, an excess of low-status men unable to find wives results in an easy (and large) pool of recruits for extremism and violence.

This shift in gender ratios may have other, less heralded implications. Some of my own work has suggested that this shift may actually shorten men's lives, reversing some of the historic progress we have made. Across a range of species, skewed sex ratios result in intensified competition for sexual partners, and this induces stress for the supernumerary sex. In humans, it seems, a 5 percent excess of males at the time of sexual maturity shortens the survival of men by about three months in late life, which is a very substantial loss.

On the other hand, the population worldwide is getting older, especially in the developed world. Globally, the United Nations estimates that the proportion of people aged sixty and over will double between 2000 and 2050, from 10 percent to 21 percent, and the proportion of children will drop from 30 percent to 21 percent. This change also has numerous implications, including for the "dependency ratio," meaning that fewer young people are available to provide for the medical and economic needs of the elderly. Much less heralded is the fact that war is a young person's activity, and it is entirely likely that as populations age, they may become less aggressive.

The changing nature of kinship networks, such as the growth in blended families—whether due to changing divorce patterns in the developed world or AIDS killing off parents in Africa—affects the network of obligations and entitlements within families. Changing kinship systems in modern American society (with complex mixtures of remarried and cohabiting couples, half-siblings, step-siblings, and so on) profoundly affect caregiving, retirement, and bequests. Who cares for Grandma? Who gets her money when she dies?

Finally, it is not just the balance between males and females, or young and old, that is changing but also the balance between rich and poor. Income inequality is reaching historic heights throughout the world. One percent of the world's people receives 57 percent of the income. Income inequality in the United States is presently at its highest recorded levels, exceeding even the Roaring Twenties. And while economic development in China has proceeded with astonishing rapidity, income is not evenly distributed; the prospects for conflict in that country as a result seem very high in the coming decades.

Since we have no real predators, a key feature of the human environment is other humans. In our rush to focus on threats such as global warming and environmental degradation, we should not ignore this fact. It is well to look around at who, and not just what, surrounds us. Population structure will change everything. Our health, wealth, and peace depend on it.

AT LAST: TECHNOLOGY WILL CHANGE EDUCATION

HAIM HARARI

HAIM HARARI is a theoretical physicist, former president of the Weizmann Institute of Science, and the author of *A View from the Eye of the Storm: Terror and Reason in the Middle East.*

Sometimes you make predictions. Sometimes you have wishful thinking. It is a pleasure to indulge in both by discussing the one and the same development that will change the world.

Today's world—its economy, industry, environment, agriculture, energy, health, food, military power, communications, you name it—is driven by knowledge. The only way to fight poverty, hunger, disease, natural catastrophes, terrorism, war, and all other evils is the creation and dissemination of knowledge: that is, research and education.

Of the 6.7 billion people on our planet, at least four billion are not participating in the knowledge revolution. Hundreds of millions are born to illiterate mothers, never drink clean water, have no medical care, never use a phone.

The buzz words of distant learning, individualized learning, and all the other technology-driven changes in education remain largely on paper, far from being a daily reality in the majority of the world's schools. The hope that affluent areas will provide remote-access good education to others has not materialized. The idea of bringing all of science, art, music, and culture to every corner of the world, and the creation of schools based on individual and group learning, teamwork, simulations, and special aids for special needs—all of these technology-enabled goals remain largely unfulfilled.

It is amazing that after decades of predictions and projections, education around the world has changed so little. Thirty years ago, pundits talked about the thoroughly computerized school. Many had fantasies regarding an entirely different structure of learning, removed from the standard traditional school-class-teacher complex that has hardly changed in the last century. It is even more amazing that no one has made significant money by applying the Information Revolution to education. With a captive consumer audience of all the schoolchildren and teachers in the world, one would think that the money made by eBay, Amazon, Google, and Facebook might be dwarfed by the profits of a very clever revolutionary idea regarding education. Yet no education-oriented company is found among the ranks of the Web billionaires.

How come the richest person on the globe is not someone who had a brilliant idea about using technology for bringing education to the billions of schoolchildren of the world? I do not know the complete answer to this question. A guess is that in other fields you can have "quickies," but not in education. The timescale of education is decades, not quarters. Another guess is that in education, you must mix the energy and creativity of the young with the wisdom and experience of the older, while in other areas the young can do it fast and without the baggage of earlier generations.

I am not necessarily bemoaning the fact that no one got into the list of richest people in the world by reforming education. But I do regret that no game-changing event has taken place on this front by exploiting what modern technology offers.

Four million Singapore citizens have a larger gross domestic product than one-hundred-thirty million Pakistanis. This is not unrelated to the miseries and problems of Pakistan, which range from poverty to terror to severe earthquake damage. The only way to change this imbalance, in the long run, is education. Nothing better can happen to the world than better education in such a country. But relying only on local efforts may take centuries. On the other hand, if Al Qaeda can reach other continents from Pakistan by using the Web, why can't the world help educate one-hundred-thirty million Pakistanis using better methods?

So, my game-changing hope and prediction is that finally something significant will change on this front. The time is ripe. A few novel ideas, aided by technologies that did not exist until recently and based on humanistic values, compassion, and a true desire to extend help to the uneducated majority of Earth's population, can do the trick.

Am I naïve, stupid, or both? Why do I think this miracle, predicted for thirty years by many and impatiently awaited by more, will happen in the coming decades?

Here are my clues:

First, a technology-driven globalization is forcing us to recognize and fear the enormous knowledge gaps between different parts of the world and between segments of society within our countries. It is a major threat to everything the world has achieved in the last hundred years, including democracy itself. Identifying the problem is an important part of the solution.

Second, the speed and price of data transmission, the advances in software systems, the feasibility of remote video interactions, the price reduction of computers, fancy screens, and other gadgets, lead to the realization that tailor-made devices for schools and education are worth designing and producing. Until now, most school computers were business computers used at school, and very few special tools were developed exclusively for education. This is beginning to change.

Third, for the first time, the generation that grew up with a computer at home is reaching the teacher ranks. The main obstacle of most education reforms has always been the training of the teachers. This should be much easier now. Recall the first generation of Americans who grew up in a car-owning family. It makes a significant difference.

Fourth, the Web-based social networks in which the children now participate pose a new challenge. The educational system must join them, because it cannot fight them. So the question is no longer "Will there be a revolution in education?" but "Will the revolution be positive or deadly?" Too many revolutions in history have led to more pain and death than to progress. We must get this one right.

Fifth, a child who comes to school with a 3G phone, iPod, or whatever, sending messages to his mother's BlackBerry and knowing in real time what is happening in the classroom of his brother or his friend miles or continents away, cannot be taught anything in the same way that I was taught. Has anyone seen lately a slide rule? A logarithmic table? A volume of Pedia other than Wiki?

At this point, I could produce long lists of specific ideas that one might try, or of small steps that have already been taken somewhere in the world. But it is unlikely that one or three or ten such ideas will do the job. It will have to be an evolutionary process, involving many innovations, trial and error, self-adjustment, avoidance of past mistakes, and above all, patience. It will also have to include one or more big game-changing elements, comparable to the influence of Google.

This is a change that will create a livable world for the next generations, both in affluent societies and, especially, in the developing or not-even-yet-developing parts of the world. Its time has definitely come. It will happen and it will, indeed, change everything.

INEXPENSIVE CUSTOMIZABLE INTERACTIVE E-TEXTS FOR WORLDWIDE USE

<o>

DAVID G. MYERS

DAVID G. MYERS is a social psychologist at Hope College, Holland, Michigan, and the author of A *Friendly Letter to Skeptics and Atheists: Musings on Why God Is Good and Faith Isn't Evil.*

My university colleagues in southern Africa have often expressed their wish for teaching materials that, for them, would change everything: If only there were a way for their students, who could not afford even greatly discounted Euro-American textbooks, to have access to low-cost, culturally relevant, state-of-the-art textbooks. For students in Africa (and elsewhere around the globe), this utopian world may, in the next decade, become the real world, thanks to the following:

- Interactive textbooks: Various publishers are developing Web-based interactive e-books with links to tutorials, simulations, quizzes, animations, virtual labs, discussion boards, and video clips. (These are not yesterday's e-textbooks.)
- Customizability: Instructors and regional instructor networks will be able to rearrange the content, delete unwanted material, and add (or link to) materials pertinent to their students' worlds and their own course goals.

- Affordability: In the developed world, students will pay for course access. With no hard-copy book production or shipping and no used books, publishers will stay afloat with a much smaller fee paid by many more students, or via a site license. For courses in economically impoverished regions, benevolent publishers could make access available for very low cost per student.
- Student accountability: Instructors will track their students' engagement in advance of class sessions, thus freeing more class time for discussion.
- Expanding broadband access: Thanks partly to a joint foundation initiative by Rockefeller, Carnegie, Ford, and others, information technologies and the Internet are coming to African universities. As yet, access is limited and expensive, but with increased bandwidth and the prospect of inexpensive wireless personal reading devices, that may change.

This is not pie in the sky. African educators are eager to explore the effectiveness of the new interactive content when it becomes available to their students. The hope is that it will combine the strengths of existing texts, which are comprehensive, expertly reviewed, painstakingly edited, and attractively packaged, and be supported with teaching aids at reduced cost and perhaps with locally adapted illustrations and content. By making the same information available to rich and poor students at rich and poor schools in rich and poor countries, e-texts are egalitarian. They flatten the world. And as James Madison noted in 1825, "The advancement and diffusion of knowledge is the only guardian of true liberty."

ON BASKETBALL AND SCIENCE CAMPS

———◇———

STEPHON H. ALEXANDER

STEPHON H. ALEXANDER is an assistant professor of physics at Penn State University.

I grew up in the northeast Bronx, where in the 1980s pretty much everyone's heroes were basketball sensations Michael Jordan and Dominique Wilkins. I and most of my friends fantasized about playing in the NBA. True, playing basketball was fun—but another obvious incentive was that aside from drug dealers, athletes were the only people from our socioeconomic background whom we saw earning serious money and respect. Despite my early tendencies toward science and math, I played hookey quite a bit, spending many hours on the P.S. 16 basketball court. There I would dream of one day making my high school basketball team and doing a 360-degree dunk. Neither happened. At fifteen, in the middle of a layup, I stumbled and broke my kneecap, which forced me off the basketball playground for half a year. I was relegated to homework and consistent class attendance.

Most of my street court pals failed to graduate from high school. But although they were far better ballplayers than I, only one made it to the NBA. A few others did get scouted and ended up playing for Big Ten basketball teams. To this day, whenever I return to my old neighborhood, I see some of my diploma-less pals doing old school moves with kneepads on.

The year of the broken knee led to a scholarship from a private donor for a summer physics camp for teens called ISI (International Summer Institute). The camp was in Southampton, Long Island, an environment far different from any I'd ever experienced. Most of the other kids were from foreign countries. I made strange new friends, including Hong, a South Korean boy who spent the summer trying to compute pi to some decimal point or other. There was a group of young chess players being coached by a Russian chess master. I took college physics. Most of my fellow campers, one of whom I am still in touch with, went on to become excellent scientists. At some point, I met the organizer of the camp, a gentleman wearing a leather jacket even in summer, who turned out to be Nobel laureate Sheldon Glashow (who coincidentally had attended my neighboring high school, the Bronx High School of Science). He gave us a physics/inspirational talk. During that talk, I realized that there are other types of Michael Jordans, in areas other than basketball, and that, like Shelley Glashow, I could be different plus make a good living as a scientist. More important, we teenagers bonded and essentially formed a young global community of future scientists.

When I returned to the Bronx, I couldn't really talk much about my experience. After all, a discussion of the Heisenberg uncertainty principle would have been far less interesting to my pals than ballpark trash talk. I began playing less basketball and eventually went on to college and became a physicist. I could not help feeling a little guilty about this. In the back of my mind, I knew the real mathematical genius in my neighborhood was a guy named Eric Deabreu. But he never finished high school.

What if there were a global organization of scientists and educators dedicated to identifying (or scouting) the potential Michael Jordans of science, regardless of what part of the world they were from and regardless of their socioeconomic background? This is happening at the local level, but not globally. What if these students were provided the resources to reach their full potential and naturally forge a global community of scientific peers and friends? What we would have, among many benefits, is an orchestrated global effort to address the most pressing scientific problems that current and future generations must confront: the energy crisis, global warming, HIV, diplomacy, to name a few. I think an initiative that markets the virtues of science on every corner of the planet with the same urgency as the basketball scouts on corners of street-ball courts would change the world. Such a reality has long been my vision. In light of the efforts of some in the science community—among them the physicists Clifford Johnson, Jim Gates, and Neil Turok—I believe I will see it come to pass.

A WEB-EMPOWERED REVOLUTION IN TEACHING

―◇―

CHRIS ANDERSON

CHRIS ANDERSON is curator of the annual TED (Technology, Entertainment, Design) Conference.

Today when we think of the world's teeming billions of humans, we tend to think of overpopulation, poverty, disease, instability, environmental destruction. They are the cause of most of the planet's problems.

What if that were to change? What if the average human were able to contribute more than consume? To add more than subtract? Think of the world as if each person drives a balance sheet. On the negative side are the resources they consume without replacing. On the positive side are the contributions they make to the planet, in the form of resources they produce, the lasting artifacts of value they build, and the ideas and technologies that might create a better future for their families, their communities, and the planet as a whole. Our future hangs on whether the sum of those balance sheets can turn positive.

What might make that possible? One key reason for hope is that so far we have barely scraped the surface of human potential. Throughout history, the vast majority of humans have not been the people they could have been.

Take this simple thought experiment. Pick your favorite scientist, mathematician, or cultural hero. Now imagine that instead of being born when and where they were, they had instead been born with the same abilities in a typical poverty-stricken village in, say, the France of 1200 or the Ethiopia of 1980. Would they have made the same contribution they did make? Of course not. They would never have received the education and encouragement it took to achieve what they did. Instead they would simply have lived out a life of poverty, with perhaps an occasional intuition that there was a better way. Conversely, an unknown but vast number of those grinding out a living today are potential world changers.

If only we could find a way of unlocking that potential. Two keys might be enough: knowledge and inspiration. If you learn how to transform your life for the better and you are inspired to act on that knowledge, there's a good chance that your life will indeed improve.

There are many scary things about today's world, but what is truly thrilling is that the means of spreading both knowledge and inspiration have never been greater. Five years ago, a teacher or professor able to change the lives of his or her students could realistically hope to reach maybe a hundred of them a year. Today that same teacher can communicate through video to millions of eager students. There are already numerous examples of powerful talks that have spread virally to massive Internet audiences. The cost of distributing a recorded lecture anywhere in the world via the Internet has effectively fallen to zero. This has happened with breathtaking speed and its implications are not yet widely understood. But it is surely capable of transforming global education.

For one thing, the realization that today's best teachers can become global celebrities will boost the caliber of those who teach. For the first time in many years, it's possible to imagine ambitious, brilliant eighteen-year-olds putting "teacher" at the top of their career choice list. Indeed, the very definition of "great teacher" will expand, as numerous people outside the profession who can communicate important ideas find a new incentive to make that talent available to the world. Additionally, teachers can amplify their own abilities by inviting into their classroom, on video, the world's greatest scientists, visionaries, and tutors.

Now, think about this from the pupils' perspective. In the past, your success depended on whether you were lucky enough to have a great mentor or teacher in your neighborhood. The vast majority have not been so fortunate. But a young girl born in Africa today will probably have access, in ten years' time, to a cell phone with a high-resolution screen, a Web connection, and more power than the computer you own today. We can imagine her obtaining face-to-face insight and encouragement from her choice of the world's great teachers. She will get a chance to be what she can be. And she might just end up being the person who saves the planet for our grandchildren.

WISDOM REBORN

———◇———

ROGER C. SCHANK

ROGER C. SCHANK is a psychologist and computer scientist, founder of the Institute for the Learning Sciences at Northwestern University, and the author of *Making Minds Less Well Educated Than Our Own*.

An executive of a consumer products company whom I know was worrying about how to improve the bleach his company produces. He thought it would be nice if the bleach didn't cause "collateral damage"; that is, he wanted it to harm bad stuff without harming good stuff. He seized upon the notion of collateral damage and began to wonder where else collateral damage was a problem. Chemotherapy came to mind, and he visited some oncologists, who gave him some ideas about what they did to make chemotherapy less harmful to patients. He then applied those ideas to improve his company's bleach.

He began to wonder about what he had done and how he had done it. He wanted to be able to do this sort of thing again. But what is "this sort of thing," and how can one do it again?

In bygone days, we lived in groups that had wise men and women who told stories to the young if they thought that those stories might be relevant to their needs. This was called "wisdom," and it served to pass one generation's experiences to the next. We have lost this ability to some extent, because we live in a much larger world, where the experts are not likely to be in the next cave over and where there is a lot more to have expertise about. Nevertheless, we as humans are set up to deliver and make use of just-in-time wisdom—though we aren't that sure where to find it. We have created books and schools and now search engines to replace what we have lost. Still, it would be nice if there were wisdom to be had without having to look hard to find it.

Those days of just-in-time storytelling will return. The storyteller will be your computer. The computers we have today are capable of understanding your needs and finding just the right (previously archived and indexed) wise man or woman to tell you a story, just when you need it, that will help you think something out. No more looking for information. No more libraries. No more key words. No more search engines. Information will find you, and just in the nick of time. And this will change everything. Some work needs to be done to make this happen, of course.

You are seeing the beginning of this today, but it is being done in a mindless and commercial way—led, of course, by Google ads that watch the words you type and match them to ads they have written that contain those words. (I receive endless offers of online degrees, for example, because that is what I often write about.) Three things will change:

1. The information that finds you will be relevant and important to what you are working on and will arrive just in time.
2. The size of information will change. No more books' worth amount of information (book size is an artifact of what-length books sell; there are no ten-page books).
3. A new form of publishing will arrive that serves to vet the information you receive. Experts will be interviewed and their best stories will be indexed. Those stories will live forever, waiting for someone to tell them to at the right moment.

In the world I am describing, the computer has to know what you are trying to accomplish, not what words you just typed, and it needs to have an enormous archive of stories to tell you. Additionally, it needs to have indexed all the stories it has in its archives to activities you are working on, in such a way that the right story comes up at the right time.

What has to happen to make this a reality? Computers need an activity model: They need to know what you are doing and why. As software becomes more complex and more responsible for what we do in our daily lives, this state of affairs is inevitable.

An archive needs to be created that has all the wisdom of the world in it. People have attempted this for years, in the form of encyclopedias and such, but they have failed to do what was necessary to make those encyclopedias useful. There is too much in a typical encyclopedia entry, not to mention the absurd amount of information in a book. People are set up to hear stories, and stories don't last all that long before we lose our ability to concentrate on their main point—their inherent wisdom, if you will. People tell each other stories all the time, but when they write or lecture they are permitted (even encouraged) to go on way too long (as I am doing now).

Wisdom depends upon goal-directed prompts that say what to do when certain conditions are encountered. To put this another way, an archive of key strategic ideas about how to achieve goals under certain conditions is just the right resource to be interacting with, enabling a good story to pop up when you need it. The solution involves goal-directed indexing. Ideas such a "collateral damage" are indices to knowledge. We are not far from the point where computers will be able to recognize collateral damage when it happens and find other examples that help you think something out.

Having a "reminding machine" that reminds us of universal wisdom as needed will indeed change everything. We will all become much more likely to profit from humanity's collective wisdom by having a computer at the ready to help us think.

TRACKS AND CLUSTERS

◦

DAVID GELERNTER

DAVID GELERNTER is a computer scientist at Yale University, chief scientist of Mirror Worlds Technologies, and the author of *Americanism: The Fourth Great Western Religion*.

What will change everything? The replacement of 90 percent of America's teachers at every level with parent-chosen, cloud-resident "learning tracks." The end of conventional centralized, age-stratified schools and their replacement by local cluster rooms, where a few dozen children of all ages and IQs gather under the supervision of any trustworthy adult and where each child follows his own learning track at his own level and rate but all kids in the cluster participate in playtime and gym-type activities together.

Thus primary and secondary education become radically localized and delocalized simultaneously. All children go to a nearby cluster room and mix there with other children of all ages and interests from the neighborhood: radical localization (or relocalization, the return of the little red schoolhouse). But each child follows a learning track prepared and presented by the best teachers and thinkers anywhere in the nation or the world. Local schools become cheap and flexible (doesn't matter whether ten children or fifty show up, as long as there are enough machines to go around—and that will be easy). Perhaps 80-plus percent of school funding goes straight to the production of learning tracks, which accumulate in a growing worldwide library.

This inversion of education has bad properties as well as good: It's much easier to learn from a good teacher face-to-face than from any kind of software. But the replacement of schools by tracks and clusters is the inevitable, unstoppable, take-it-or-leave-it response to educational collapse in the United States. The National Commission on Excellence in Education's report, *A Nation at Risk*, appeared in 1983. Americans have known for a full generation that their schools are collapsing and have failed even to make a dent in the problem. If anything, today's schools are worse than 1983's. Tracks-and-clusters is no perfect solution—but radical change is coming, and cloud-based, parent-chosen tracks with local cluster rooms are all but inevitable.

None of today's software frameworks for online learning is adequate. New software must make it easy to see and evaluate each track as a whole, give learners control over learning, integrate multimedia smoothly, include students in a Net-wide discussion of each topic, and put them in touch with (human) teachers as needed. New software must also make it easy for parents and "guidance teachers" to evaluate each child's progress. It's all easily done with current technology—*if* software design is taken seriously.

Any person or group can offer a learning track at any level, on any topic. The usual consumer evaluation mechanisms will help parents and students choose. Government and private organizations will review learning tracks, comment, and mark them "approved" or not. Suggested curricula will proliferate on the Net. Anybody will be free to offer his or her services as a personal learning consultant.

Tracks-and-clusters poses many problems—and suggests many solutions. It represents the inevitable direction of education in the United States, not because it solves every problem but because the current system is intellectually bankrupt—not merely today's schools and school districts, but the whole system of government funding, local school boards and budget votes, approved textbooks, nationwide educational fads, and so on. They're all ripe for the trash and on their way out. U.S. schools will change radically, because—and only because—they *must* change radically. Ten years from now, the move to clusters-and-tracks will be well under way.

THE MOBILE PHONE

———◦———

KEITH DEVLIN

KEITH DEVLIN is a mathematician, executive director of
the Center for the Study of Language and Information at
Stanford, and the author of *The Unfinished Game: Pascal,
Fermat, and the Seventeenth-Century Letter That Made the
World Modern.*

This is a tough one. Not because there is a shortage of possibilities for major advances in science, and not because any predictions Edgies make are likely to be way off the mark (history
tells us that they assuredly will be); rather, John Brockman has
set the hurdle impossibly high with "change everything" and "expect to live to see." The contraceptive pill "changed everything"
for people living in parts of the world where it's available, and
the Internet "changed everything" for those of us who are connected. But for large parts of the world, those advances may as
well not have occurred. Moreover, many scientific changes take
a generation or more to have a significant effect.

But since I'm asked, I'll come up with an answer, and it's
one I am pretty sure will happen in my lifetime (say, thirty
more years). The reason for my confidence? The key scientific
and technological steps have already been taken. In giving my
answer, I'm adopting a somewhat lawyerlike strategy of taking
advantage of that word "developments" in the *Edge* question.
Scientific advances do not take place purely in the laboratory,

particularly game-changing ones. They have to find their way into society as a whole, and that transition is an integral part of any "scientific advance."

History tells us that it can often take some time for a scientific or technological advance to truly "change everything." Electricity, the lightbulb, the internal combustion engine, and the relationship between germs and disease are classic examples. (Even these still have not affected everyone on the planet, of course—at least not directly, but that is surely just a matter of time.) The development I'll focus on is the final one in the scientific chain that brings the results of the science into everyday use.

My answer? It's staring us in the face: the mobile phone. Within my lifetime, I fully expect almost every living human adult, and most children, in the world to own one. (Neither the pen nor the typewriter came even close to that level of adoption, nor has the automobile.) The mobile phone puts global connectivity, immense computational power, and access to all the world's knowledge amassed over many centuries, into everyone's hands. The world has never, ever, been in that situation before. It really will change everything, from the way individual people live their lives to the way wealth and power are spread across the globe. It is the ultimate democratizing technology. And if my answer seems less "cutting-edge" or scientifically sexy than many of the others discussed here, that just shows how dramatic and pervasive the change has already been.

What other object do you habitually carry around with you and use all the time and take for granted? Yet when did you acquire your first mobile phone? Can you think of a reason why anyone else in the world will not react the same way when the technology reaches them? Now imagine the effect on someone in a part of the world that has not had telephones, computers, the Internet, or even easy access to libraries. I'll let your own answers to these questions support my case that this is game changing on a hitherto unknown global scale.

ENERGY AND ECONOMICS: THE ROAD TO CIVILIZATION 1.0

MICHAEL SHERMER

MICHAEL SHERMER is the publisher of *Skeptic* magazine, a monthly columnist for *Scientific American,* and the
author of *The Mind of the Market: Compassionate Apes,
Competitive Humans, and Other Tales from Evolutionary
Economics.*

This year finds us at a crisis tipping point, both economically
and environmentally. If ever we needed to look to the past to save
our future, it is now. In particular, we need to do two things: (1)
Stop the implosion of the economy and enable markets to function once again both freely and fairly, and (2) make the transition from nonrenewable fossil fuels as the primary source of our
energy to renewable energy sources that will allow us to flourish
into the future. Failure to make these transformations will doom
us to the endless tribal political machinations and economic
conflicts that have plagued civilization for millennia. We need
to make the transition to Civilization 1.0. Let me explain.

In a 1964 article on searching for extraterrestrial civilizations, the Soviet astronomer Nikolai Kardashev suggested using
radio telescopes to detect energy signals from other solar systems
in which there might be civilizations of three levels of advancement: Type I can harness all of the energy of its home planet,
Type II can harvest all of the power of its sun, and Type III can
master all of the energy from its galaxy.

Based on our energy efficiency at the time, in 1973 the astronomer Carl Sagan estimated that Earth represented a Type 0.7 civilization, on a Type 0 to Type 1 scale. (Later assessments have put us at 0.72.) As the Kardashevian scale is logarithmic, where any increase in power consumption requires a huge leap in power production, fossil fuels won't get us to Civilization 1.0. Renewable sources, such as solar, wind, and geothermal, are a good start. Coupled to nuclear power—perhaps even nuclear fusion, instead of the fission reactors we have now—they could eventually get us there.

We are close. Taking a Janus-faced look to the past in order to see the future, let's quickly review the history of humanity on its climb to become a Civilization 1.0:

TYPE 0.1: Fluid groups of hominids living in Africa. Technology consists of primitive stone tools. Intragroup conflicts are resolved through dominance hierarchy, and intergroup violence is common.

TYPE 0.2: Bands of roaming hunter-gatherers that form kinship groups with a mostly horizontal political system and an egalitarian economy.

TYPE 0.3: Tribes of individuals linked through kinship but with a more settled and agrarian lifestyle. The beginnings of a political hierarchy and a primitive economic division of labor.

TYPE 0.4: Chiefdoms consisting of a coalition of tribes into a single hierarchical political unit with a dominant leader at the top, and with the beginnings of significant economic inequalities and a division of labor in which lower-class members provide food and other products consumed by nonproducing upper-class members.

TYPE 0.5: The state as a political coalition with jurisdiction over a well-defined territory and its inhabitants, with a mercantile economy that seeks a favorable balance of trade in a zero-sum game against other states.

TYPE 0.6: Empires extend control over peoples who are not culturally, ethnically, or geographically within their normal jurisdiction, with a goal of economic dominance over rival empires.

TYPE 0.7: Democracies that divide power among several institutions, which are run by elected officials voted for by some citizens. The beginnings of a market economy.

TYPE 0.8: Liberal democracies that give the vote to all citizens. Markets that begin to embrace a nonzero, win-win economic game through free trade with other states.

TYPE 0.9: Democratic capitalism, the blending of liberal democracy and free markets, now spreading across the globe through democratic movements in developing nations and broad trading blocs such as the European Union.

TYPE 1.0: Globalism that includes worldwide wireless Internet access, with all knowledge digitized and available to everyone. A global economy with free markets in which anyone can trade with anyone else without interference from states or governments. A planet where all states are democracies in which everyone has the franchise.

Looking from this past toward the future, we can see that the forces at work that could prevent us from reaching Civilization 1.0 are primarily political and economic, not technological. The resistance by nondemocratic states to turning power over to the people is considerable, especially in theocracies whose leaders would prefer we all revert to Type 0.4 chiefdoms. The opposition to a global economy is substantial, even in the industrialized West, where economic tribalism still dominates the thinking of most people.

The game-changing scientific idea is the combination of energy and economics—the development of renewable energy sources made cheap and available to everyone, everywhere on the planet, by allowing anyone to trade in these game-changing technologies with anyone else. That will change everything.

UNDOING BABYLON

———◦———

DANIEL L. EVERETT

DANIEL L. EVERETT is chair of the Department of Languages, Literatures, and Cultures and professor of linguistics and anthropology at Illinois State University and the author of *Don't Sleep, There Are Snakes: Life and Language in the Amazon Jungle*.

> "We should really not be studying sentences; we should not be studying language—we should be studying people."
>
> —VICTOR YNGVE

Communication is the key to cooperation. Although cross-cultural communication for the masses requires translation techniques that exceed our current capabilities, the groundwork of this technology has already been laid, and many of us will live to see a revolution in automatic translation that will change everything about cooperation and communication across the world.

This goal was conceived in the late 1940s in a famous memorandum by Rockefeller Foundation scientist Warren Weaver, in which he suggested the possibility of machine translation and tied its likelihood to four proposals, still controversial today: that there was a common logic to languages; that there were likely to be language universals; that immediate context could be understood and linked to translation of individual sentences; and that cryptographic methods developed in World War II would apply to language translation. Weaver's proposals got off the ground financially in the early 1950s as the U.S. military invested heavily in linguistics and machine translation across the United States, with particular emphasis on the research of the team headed by Victor Yngve at the Massachusetts Institute of Technology's Research Laboratory of Electronics—a team that included the young Noam Chomsky.

Yngve, like Weaver, wanted to contribute to international understanding by applying the methods of computational linguistics, the then-incipient field he helped found, to communication, especially machine translation. Early innovators in this area also included Claude Shannon at Bell Labs and Yehoshua Bar-Hillel, who preceded Yngve at MIT before returning to Israel. Shannon was arguably the inventor of the concept of information as an entity that could be scientifically studied, and Bar-Hillel was the first person to work full-time on machine translation, beginning the program that Yngve inherited at MIT.

This project was challenged early on, however, by the work of Chomsky, from within Yngve's own lab. Chomsky's conclusions about different grammar types and their relative generative power convinced people that grammars of natural languages were not amenable to machine-translation efforts as they were practiced at the time, leading to a slowdown in and reduction of enthusiasm for computationally based translation.

As we have subsequently learned, however, the principal problem faced in machine translation is not the formalization of grammar per se, but the inability of any formalization known, including Chomsky's, to integrate context and culture (semantics and pragmatics, in particular) into a model of language appropriate for translation. Without this integration, mechanical translation from one language to another is not possible.

Still, mechanical procedures able to translate most contents from any source language into accurate, idiomatically natural constructions of any target language seem less utopian to us now, because of major breakthroughs that have led to several programs in machine translation (for example, the Language Technologies Institute at Carnegie Mellon University). I believe we will see within our lifetime the convergence of developments in artificial intelligence, knowledge representation, statistical grammar theories, and the emerging field of computational anthropology (informatic-based analysis and modeling of cultural values) that will facilitate powerful new forms of machine translation to match the dreams of early pioneers of computation.

The conceptual breakthroughs necessary for universal machine translation will also require contributions from construction grammars—models that view language as a set of conventional signs (varieties of the idea that the building blocks of grammar are not rules or formal constraints, but conventional phrase and word forms that combine cultural values and grammatical principles) instead of as a list of formal properties. These breakthroughs will have to look at differences in the encoding of language and culture across communities rather than trying to find a "universal grammar" that unites all languages.

At least some of the steps are easy enough to imagine. First, we come up with a standard format for writing statistically based construction grammars of any language, a format that displays the connections between constructions, culture, and local context (such as the other likely words in the sentence, or other likely sentences in the paragraph, in which the construction appears). This format might be as simple as a flowchart or a list. Second, we develop a method for encoding context and values. For example, what are the values associated with words; what are the values associated with certain idioms; what are the values associated with the ways in which ideas are expressed? This last can be seen in the notion of sentence complexity, for example, as in (among others) the Amazon Pirahãs' rejection of recursive structures in syntax because they violate principles of information rate and new versus old information in utterances that are very important in Pirahã culture. Third, we establish lists of cultural values and most common contexts and how these link to individual constructions. Automating the procedure for discovering or enumerating these links will take us to the threshold of automatic translation in the original sense.

Information and its exchange form the soul of human cultures. So just imagine the possible change in our perceptions of "others" when we are able to type in a story and have it automatically and idiomatically translated with 100 percent accuracy into any language for which we have a grammar of constructions. Imagine speaking into a microphone and having your words come out in the language of your audience, heard and understood naturally. Imagine being able to take a course in any language, from any university in the world, over the Internet or in person, without having to first learn the language of the instructor.

These will always be unreachable goals to some degree. It seems unlikely, for example, that all grammars and cultures are even capable of expressing everything from all languages. However, we are developing tools that will dramatically narrow the gaps and help us decide where and how we can communicate particular ideas cross-culturally. Success at machine translation might not end all the world's sociocultural or political tensions, but it won't hurt. One struggles to think of a greater contribution to world cooperation than progress to universal communication, enabling all and sundry to communicate with nearly all and sundry. Babel means "the gate of god." In the Bible, it is about the origin of world competition and suspicion. As humans approached the entrance to divine power by means of their universal cooperation via universal communication, so the biblical story goes, language diversity was introduced to destroy our unity and deprive us of our full potential.

But automated, near-universal translation is coming. And it will change everything.

SOUL TRAVEL FOR SELFLESS BEINGS

———◇———

THOMAS METZINGER

THOMAS METZINGER is a professor of philosophy at the Johannes Gutenberg-Universität Mainz and the author of *The Ego Tunnel: The Science of the Mind and the Myth of the Self.*

John Brockman points out that new technology leads not only to new ways of perceiving ourselves, but also to a process he calls "re-creating ourselves." Could this become true in an even deeper and more radical way than through gene technology? The answer is yes.

It is entirely plausible that we may one day directly control virtual models of our own bodies directly with our brain. In 2007, I first experienced taking control of a computer-generated whole-body model of myself. It took place in a virtual-reality lab, where my physical motions were filmed by eighteen cameras picking up signals from sensors attached to my body. Over the past two years, research groups in Switzerland, England, Germany, and Sweden have demonstrated how, in a passive condition, subjects can consciously *identify* with the content of a computer-generated virtual-body representation, fully relocating the phenomenal sense of self into an artificial, visual model of their body.

In 2008, in another experiment, we saw that a monkey on a treadmill could control the real-time walking patterns of a humanoid robot via a brain-machine interface directly implanted into the monkey's brain. The synchronized robot was in Japan, while the poor monkey was located thousands of miles away, in the United States. Even after it stopped walking, the monkey was able to sustain the locomotion of the synchronized robot for a few more minutes, just by using the visual feedback transmitted from Japan plus its own "thoughts" (whatever those may have been).

Now imagine two further steps.

First, we manage to selectively block the high-bandwidth "interoceptive" input into the human self–model—all the gut feelings and the incessant flow of inner body perceptions that anchor the conscious self in the physical body. After all, we already have selective motor control for an artificial body-model and robust phenomenal self-identification via touch and sight. By blocking the internal self-perception of the body, we could be able to suspend the persistent causal link to the physical body.

Second, we develop richer and more complex avatars, virtual agents emulating not only the proprioceptive feedback generated by situated movement, but also certain abstract aspects of ongoing global control itself—new tools, as Brockman would call them. Then suddenly it happens that the functional core process initiating the complex control loop connecting physical and virtual body jumps from the biological brain into the avatar.

I don't believe this will happen tomorrow. I also don't believe that it will change everything. But it will change a lot.

INSIDE OUT: THE EPISTEMOLOGY OF EVERYTHING

———◇———

TOR NØRRETRANDERS

TOR NØRRETRANDERS is a writer and lecturer on popular science and the author of *The Generous Man: How Helping Others Is the Sexiest Thing You Can Do*.

Understanding that the outside world is really inside us and the inside world is really outside us will change everything, both inside and outside. Why?

"There is no *out there* out there," physicist John Wheeler said, in his attempt to explain quantum physics. All we know is how we correlate with the world. We do not really know what the world is *really* like, uncorrelated with us. When we seem to experience an external world that is out there independent of us, it is something we dream up.

Modern neurobiology has reached the exact same conclusion. The visual world—what we see—is an illusion, but a very sophisticated one. There are no colors, no tones, no constancy in the "real" world; it is all something we make up. We do so for good reasons and with great survival value. Because colors, tones, and constancy are expressions of how we correlate with the world.

The merging of the epistemological lesson from quantum mechanics with the epistemological lesson from neurobiology attests to a very simple fact: What we perceive as being outside us is a fancy and elegant projection of what we have inside. We make this projection as a result of interacting with something that is not inside us, but everything we experience is inside.

Is this experience not real? It embodies a correlation that is very real. As physicist N. David Mermin has argued, we have correlations, but we don't know what it is that correlates, or if any correlata exist at all. This is a modern formulation of quantum pioneer Niels Bohr's view that "physics is not about nature, it is about what we can say about nature."

So what is real, then? Inside us humans, a lot of relational emotions exist. We feel affection, awe, warmth, glow, mania, belonging, and repulsion—toward other humans and toward the world as a whole. We relate, and that provokes deep emotional states. These are real and true. Being inside our bodies, these states are perceived not as "real states" of the outside world but more like a kind of weather phenomenon inside us.

That raises the simple question: Where do these internal states come from? Did we make them, or did they make us? Love existed before us (most of us were conceived in an act of love). Friendship, family bonds, hate, anger, trust, distrust—all of these entities existed before the individual. They are primary. The illusion of the ego denies the fact that they are there before the ego consciously decided to love or hate or care or not. But the inner states predate the conscious ego. And they predate the bodily individual.

The emotional states inside us are very real and the product of biological evolution. They are helpful to us in our attempt to survive. Experimental economics and behavioral sciences have recently shown us how important they are to us as social creatures: To cooperate, you have to trust the other party, even though a rational analysis will tell you that both the likelihood and the cost of being cheated is very high. When you trust, you experience a physiologically detectable inner glow of pleasure. So the inner emotional state says yes. However, if you rationally consider the objects in the outside world, the other parties, and consider their trade-offs and motives, you ought to choose not to cooperate. Analyzing the outside world makes you say no. Human cooperation is dependent on our giving weight to what we experience as the inner world compared to what we experience as the outer world.

Traditionally, the culture of science has denied the relevance of the inner states. Now they have become increasingly important to understanding humans, and highly relevant when we want to build artifacts that mimic us.

Soon we will be building not just artificial intelligence but also artificial will: systems able to convert internal decisions and values into external change. They will be able to decide they want to change the world. A plan inside becomes an action on the outside. So they will have to know what is inside and what is outside.

In building these machines, we will learn something that will change everything: The trick of perception is the trick of mistaking an inner world for the outside world. The emotions inside are the evolutionary reality. The things we see and hear outside are just elegant ways of imagining correlata that can explain our emotions, our correlations. We don't hear the croak, we hear the frog. When we understand that the inner emotional states are more real than what we experience as the outside world, cooperation becomes easier. The epoch of insane mania for rational control will be over.

What really changes is the way we see things, the way we experience things. For anything to change out there, you have to change everything in here. That is the epistemological situation. All spiritual traditions have been talking about it. But now it grows from the epistemology of quantum physics, neurobiology, and the building of robots.

We will be sitting there, building those artificial-will robots. Suddenly we will start laughing. There is no *out there* out there; it is in here. There is no *in here* in here; it is out there. The outside is in here. Who is there?

That laughter will change everything.

CHANGES IN THE CHANGERS

———◦———

A. GARRETT LISI

A. GARRETT LISI is an independent theoretical physicist and the author of "An Exceptionally Simple Theory of Everything."

Human beings have an amazingly flexible sense of self. If we don a pair of high-resolution goggles showing the point of view from another body, with feedback and control, we perceive ourselves to be that body. As we use rudimentary or complex tools, these quickly become familiar extensions of our bodies and minds. This flexibility, and our indefatigable drive to learn, invent, have fun, and seek new adventure, will lead us down future paths that will dramatically alter human experience and our very nature.

Because we adapt so quickly, the changes will feel gradual. In the next few years, solid-state memory will replace hard drives, removing the mechanical barrier to miniaturization of our computational gadgetry. Battery size remains a barrier to progress, but this will improve, along with increased efficiency of our electronics, and we will live with pervasive computational presence. Privacy will vanish. People will record and share their sensorium feeds with the world, and the world will share experiences. Every physical location will be geotagged with an overlay of information. Cities will become more pleasant as the internal combustion engine is replaced with silent electric vehicles that don't belch toxic fumes. We'll be drawn into the ever-evolving and persistently available conversations among our social networks. Primitive EEGs will be replaced by magnetoencephalography and functional MRI, backed by the computational power to recognize our active thought patterns and translate them to transmittable words, images, and actions. Our friends and family who wish it, and our entire external and internal world, will be reachable with our thoughts. This augmentation will change what it means to be human. Many people will turn away from their meat existence to virtual worlds, which they will fill with meaning—spending time working on science, virtual constructions, socializing, or just playing games. And we humans will create others like us but not.

Synthetic intelligence will arrive, but slowly, and it will be different enough that many won't acknowledge it for what it is. People used to think that a computer mastering chess, voice recognition, and automated driving would signal the arrival of artificial intelligence. But these milestones have been achieved, and we now consider them the result of brute-force computation and clever coding rather than bellwethers of synthetic intelligence. Similarly, computers are just becoming able to play the game of Go at the *dan* level and will soon surpass the best human players. They will pass Turing's test. But this synthetic intelligence, however adaptable, is inhuman and foreign, and many people won't accept it as more than number crunching and good programming. A more interesting sign that synthetic intelligence has arrived will be when CAPTCHAs (for Completely Automated Public Turing test to tell Computers and Humans Apart) and reverse Turing tests appear that exclude humans. The computers will have a good laugh about that. If it doesn't happen earlier, this level of AI will arrive once computers achieve the computational power to run real-time simulations of an entire human brain. Shortly after that, we will no longer be the game changers. But by then humans may have significantly altered themselves through direct biological manipulation.

The change I expect to see that will most affect human existence will come from biohacking: tissue engineering, purposefully altering genomes, and other advances in biology. Humans are haphazardly assembled biological machines. Our DNA was written by monkeys banging away, not at typewriters, but at one another, for millions of years. Imagine how quickly life will transform when DNA and biochemistry are altered with thoughtful intent. Nanotechnology already exists as the machinery within our own biological cells; we're just now learning how these machines work and how to control them. Pharmaceuticals will be customized to match our personal genome. We're going to be designing and growing organisms to suit our purposes. These organisms will sequester carbon, process raw material, and eventually repair and replace our own bodies.

It may not happen within my lifetime, but the biggest game change will be the ultimate synthesis of computation and biology. Biotech will eventually allow our brains to be scanned at a level sufficient to preserve our memories and reproduce our consciousness when uploaded to a more efficient computational substrate. At this point, our minds may be copied and, if desired, embedded and connected to the somatic helms of designed biological forms. We will become branching selves, following many different paths at once for the adventure, the fun, and the love of it. Life in the real world presents extremely rich experiences, and uploaded intelligences in virtual worlds will come outside, where they can fly as a falcon, sprint as a cheetah, love, play, or even just breathe with superhuman consciousness, no lag, and infinite bandwidth. People will dance with nature in all its possible forms. And we'll kitesurf.

Kitesurfing, you see, is a hell of a lot of fun—and kites are the future of sailing. Even though the sport is only a few years old and kite design is not yet mature, kitesurfers have recently broken the world sailing speed record, reaching over fifty knots. Many in the sailing world are resisting the change and disputing the record, but kites provide efficient power and lift, and the speed gap will only grow as technology improves. Kitesurfing is a challenging dynamic balance of powerful natural forces. It feels wonderful and it's even more fun in waves.

All of these predicted changes are extrapolations from the seeds of present science and technology. The biggest surprises will come from what can't be extrapolated. It is uncertain how many of these changes will happen within our lifetimes, because that timescale is a dependent variable and life is uncertain. It is both incredibly tragic and fantastically inspiring that our generation may be the last to die of old age. If extending our lives eludes us, cryonics exists as a stopgap gamble—Pascal's wager for Singularitarians, with an uncertain future preferable to a certain lack of one. And if I'm wrong about these predictions, death will mean I'll never know.

NEUROCOSMETICS

———◇———

MARCEL KINSBOURNE

MARCEL KINSBOURNE is a neurologist and cognitive neuroscientist at The New School and the coauthor, with Paula J. Caplan, of *Children's Learning and Attention Problems*.

"If it were possible to become free of negative emotion by a riskless implantation of an electrode—without impairing intelligence and the critical mind—I would be the first patient."

—THE DALAI LAMA, ADDRESS,
"ON THE NEUROSCIENCE OF MEDITATION,"
2005 SOCIETY OF NEUROSCIENCE ANNUAL MEETING.

Innovation in science and technology will continue to bring much change. But since it is the brain that experiences change, only changing the brain itself can possibly change everything.

Changing the human brain is not new, when it is a matter of correcting psychopathology. But whether the usual agents—psychoactive drugs, psychosurgery, electroshock, even what we eat, drink, and smoke—can change a brain that is functioning normally (that is, other than for the worse) is not known. However, the novel method of deep brain stimulation (DBS), by which electrodes are inserted into the brain to electrically stimulate precisely specified locations, is already used to correct certain brain disorders, among them Parkinsonism and obsessive-compulsive disorder.

The targeted symptoms are often relieved, but there have been profound changes in personality although the prior personality was not abnormal. A patient of lifelong somber disposition may not only be relieved of obsessions but also shift to a cheerful mood the instant the current is switched on, and revert to his subdued self the instant it is switched off. The half-empty glass temporarily becomes the glass half full. The brain seems not entirely to respect our conventional distinction between what is normal and what is not.

The stimulation's immediate effect is shocking. We assume that our gratifyingly complex minds and brains are incrementally shaped by innumerable dynamic and environmental factors. And yet identical twins, adopted apart into sharply contrasting social and economic environments, have shown impressive similarities in mood and sense of well-being. Genetically determined types of brain organization appear to set the emotional tone; experience modulates it in a positive or negative direction. Stimulating or disrupting neural transmission along a specific neural pathway or reverberating loop may reset the emotional tone, sidestepping the complexities of early experience, stress, and misfortune and letting personality float free.

The procedure releases a previously unsuspected potential. The human brain is famously plastic. Adjusting key circuitry presumably has wide repercussions throughout the brain's neural network, which settles into a different state. We have yet to thoroughly digest the philosophical implications, but a more unequivocal validation of psychoneural identity theory (the identity of the brain and the mind, different aspects of the same thing) can hardly be imagined.

Certainly, deep brain stimulation is not currently used to render sane people more thoughtful, agreeable, gentle, or considerate. Beyond the potential adverse neurosurgical side effects, there are ethical considerations that prohibit using DBS to enhance a brain considered normal. But history teaches two lessons: Any technology will tend to become more precise, effective, and safer over time; and anything that can be done ultimately will be done, philosophical and ethical considerations notwithstanding.

The example of cosmetic plastic surgery is instructive. Reconstructive in its origins, it is increasingly used for cosmetic purposes. I predict the same shift for deep brain stimulation. Cosmetic surgery is used to render people more appealing. In human affairs, appearance is critical. For our hypersocial species, personal appeal opens doors that remain shut to mere competence and intellect. Undoubtedly cosmetic surgery enhances quality of life, so how can it be denied to anyone? And yet it is, by its very nature, deceptive; the operated face is not really the person's face, the operated body not really the person's body. However, experience teaches that these reservations as to authenticity remain theoretical. The cosmetically adjusted nose, breast, thighs, or skin tones become the person's new reality, without significant social backlash. Even face transplants are now feasible. We read so much into a face—but what if it is not the person's "real" face? Does anyone care, or even remember the person's previous appearance? So it will be with neurocosmetics.

And yet is it not more deeply disturbing to tinker with the brain than to adjust one's body to one's liking? It is. However, the mind-body distinction has become somewhat blurred of late. Evidence accumulates as to the embodiment of cognition and emotion, and at the least there is influential feedback between the two domains. Considerations that will be raised by cosmetic DBS are already in play, in a minor key, with cosmetic surgery.

Deep brain stimulation seems not to enhance intellect, but intellect is no high road to success. "Social intelligence" is of prime importance, and it is a by-product of personality. Within my lifetime, DBS in some form will be used to modify personality so as to optimize professional and social opportunity. Ethicists will deplore this, and so they should. But it will happen nonetheless, and it will change in as yet unimagined ways how humans experience the world and how they relate to one another.

Consider an arms race in affability, a competition based not on concealing real feelings but on feelings engineered to be real. Consider a society of homogenized goodwill, whose citizens make regular visits to the DBS provider who advertises superior electrode placement. Switching a personality on—and then off when it becomes boring. Alternating personalities: Dr. Accumbens and Mr. Insula (friendly and disgusted, respectively). Tracking fashion trends in personality. Coordinating personalities for special events. Demanding personalities such as emerge on drugs (for example, cocaine) or in psychopathologies (for example, hypomania). Regardless, the beneficiaries of deep brain stimulation will experience life quite differently. Employment opportunities for yet more ethicists and more philosophers!

We take ourselves to be durable minds in stable bodies. But this reassuring self-concept will turn out to be yet another of our so-human egocentric delusions. Do we, strictly speaking, own stable identities? When it sinks in that the continuity of our experience of the world and of our self is at the whim of an electrical current, then our fantasies of permanence will have yielded to the reality of our fragile and ephemeral identities.

NEUROPHENOMICS + TARGETED STIMULATION = PSYCHOLOGICAL OPTIMIZATION?

———◄○►———

BRIAN KNUTSON

BRIAN KNUTSON is an associate professor of psychology and neuroscience at Stanford University.

The fashionable phrase "game changing" can imply not only winning a game (usually with a dramatic turnaround), but also changing the rules of the game. If we could change the rules of the mind, we would alter our perception of the world, which would change everything (at least for humans).

Assuming that the brain is the organ of the mind, what are the brain's rules, and how might we transcend them? Technological developments that combine neurophenomics with targeted stimulation will offer answers within the next century. In contrast to genomics, less talk (and funding) has been directed toward phenomics—the study of phenotypes, or the observable characteristics of organisms. Yet phenomics is the logical endpoint of genomics (and a potential bottleneck for clinical applications). Phenomics has traditionally focused on a broad range of individual characteristics, including morphology, biochemistry, physiology, and behavior. "Neurophenomics," however, might more specifically focus on patterns of brain activity that generate behavior.

Advances in brain-imaging techniques over the past two decades now allow scientists to visualize changes in the activity of deep-seated brain regions at a spatial resolution of less than a millimeter and a temporal resolution of less than a second. These technological breakthroughs have sparked an interdisciplinary revolution that will culminate in the mapping of a "neurophenome." The neural patterns of activity that make up the neurophenome may have genetic and epigenetic underpinnings but can also respond dynamically to environmental contingencies. The neurophenome should link more closely than behavior to the genome, could have one-to-many or many-to-one mappings to behavior, and might ideally explain why groups of genes and behaviors tend to travel together.

Although mapping the neurophenome might sound like a hopelessly complex scientific challenge, emerging research has begun to reveal a number of neural signatures that reliably index not only the obvious starting targets of sensory input and motor output but also more abstract mental constructs, like anticipation of gain, anticipation of loss, self-reflection, conflict between choices, impulse inhibition, and memory storage/retrieval (to name but a few). By triangulating across different brain-imaging modalities, the neurophenome will eventually point us toward spatially, temporally, and chemically specific targets for stimulation.

Targeted neural stimulation has been possible for decades, starting with electrical methods and followed by chemical methods. Unfortunately, delivery of any signal to deep brain regions is usually invasive (for example, requiring drilling holes in the skull and implanting wires, or worse), unspecific (for example, requiring infusion of neurotransmitters over minutes to distributed regions), and often transient (for example, target structures die or protective structures coat foreign probes). Fortunately, better methods are on the horizon. In addition to developing ever smaller and more temporally precise electrical and chemical delivery devices, scientists can now nearly instantaneously increase or decrease the firing of specific neurons with light probes that activate photosensitive ion channels. As with the electrical and chemical probes, these light probes can be inserted into the brains of living animals and change ongoing behavior. But at present, scientists still have to insert invasive probes into the brain. What if one could deliver the same spatially and temporally targeted bolus of electricity, chemistry, or even light to a specific brain location without opening the skull? Such technology does not yet exist, but given the creativity, brilliance, and pace of recent scientific advances, I expect that relevant tools will emerge in the next decade (imagine the market for triangulation helmets). Targeted and noninvasive stimulation, combined with the map that comprises the neurophenome, will revolutionize our ability to control our minds.

Clinical implications of this type of control are straightforward, yet startling. Both psychotherapy and pharmacotherapy look like blunt instruments by comparison. Imagine giving doctors or even patients the ability to precisely and dynamically control the firing of acetylcholine neurons in the case of dementia, dopamine neurons in the case of Parkinson's disease, or serotonin neurons in the case of unipolar depression. These and similar technological developments will not only improve clinical treatment but also advance scientific theory. Along with applications designed to cure will come demands for applications designed to enhance. What if we could precisely but noninvasively modulate mood, alertness, memory, control, willpower, and more? Of course, everyone wants to win the brain game. But are we ready for the rules to change?

CELEBRATORY SELF-REENGINEERING

————◇————

ANDY CLARK

ANDY CLARK holds the Chair in Logic and Metaphysics at the University of Edinburgh and is the author of *Supersizing the Mind: Embodiment, Action, and Cognitive Extension*.

What will change everything is the onset of celebratory species self-reengineering.

The technologies are pouring in, from wearable, implantable, and pervasive computing, to the radical feature blends achieved using gene-transfer techniques, to thought-controlled cursors freeing victims of locked-in syndrome, to funkier prosthetic legs able to win track races, and on to the humble but transformative iPhone.

But what really matters is the way we are just starting to know ourselves as a result of this tidal wave of self-reengineering opportunity—not as firmly bounded biological organisms but as delightfully reconfigurable nodes in a flux of information, communication, and action. As we learn to celebrate our own potential, we will embrace ever more dramatic variations in bodily form and in our effective cognitive profiles. The humans of the next century will be vastly more heterogeneous, more varied along physical and cognitive dimensions, than those of the past, as we deliberately engineer a new Cambrian explosion of body and mind.

A DIFFERENT KIND OF MALE SUBJECTIVITY

———◇———

TINO SEHGAL

TINO SEHGAL is an artist.

I think what I will live to see is a quite different kind of male subjectivity. As we increasingly shed the cultural and behavioral reflexes of industrial society, with its distinctive separation of labor between men and women, the base for our still-pervasive idea of what constitutes "masculinity" is equally eroding. Eventually this will trickle down to the concept of male subjectivity, with each new generation making its little step.

I hope I will live to see this, as the codes and modes of behavior and expression available to men are extremely limited and simplistic. I am looking forward to seeing my child grow up and learning what his generation's contribution will be—but even more so, I am impatient to see how young men will be when I am very old.

HIDDEN PERSUADERS '09

———◦———

HELEN FISHER

HELEN FISHER is a research professor in the Department of Anthropology at Rutgers University and the author of *Why We Love: The Nature and Chemistry of Romantic Love.*

"Mind is primarily a verb," wrote philosopher John Dewey. Every time we do or think or feel anything, the brain is doing something. But what? And can we use what scientists are learning about these neural gymnastics to get what we want? I think we can and we will, in my lifetime, thanks to some mind-bending developments in contemporary neuroscience. Brain scanning, genetic studies, antidepressant-drug use, estrogen-replacement therapy, testosterone patches, L-dopa and newer drugs to prevent or retard brain diseases, recreational drugs, sex-change patients, gene doping by athletes—all these and other developments are giving us data on how the mind works and opening new avenues to use brain chemistry to change who we are and what we want. As the field of epigenetics takes on speed, we are also beginning to understand how the environment affects brain systems and even turns genes on and off, further enabling us to adjust brain chemistry, affecting who we are, how we feel, and what we think we need.

But is this new? Our forebears have been manipulating brain chemistry for millions of years. Take "hooking up," the current version of the one-night stand—among humankind's oldest forms of chemical persuasion. During sex, stimulation of the genitals escalates activity in the dopamine system, the neurotransmitter network my colleagues and I have found to be associated with feelings of romantic love. And during orgasm you experience a flood of oxytocin and vasopressin, neurochemicals associated with feelings of attachment. Casual sex isn't always casual, and I suspect that our ancestors seduced one another in order (unconsciously) to alter their partner's brain chemistry, thereby nudging him or her toward feelings of passion and/or attachment. Indeed, this chemical persuasion works. In a recent study of five-hundred-seven college students, anthropologist Justin Garcia found that 50 percent of the women and 52 percent of the men hopped into bed with an acquaintance or a stranger in hopes of starting a longer relationship, and about a third of these hook-ups turned into romance.

In 1957, Vance Packard wrote *The Hidden Persuaders* to unmask the subtle psychological techniques advertisers use to manipulate people's feelings and induce them to buy. We have long been using psychology to persuade others, but now we are learning why our psychological strategies work. Holding hands, for example, generates feelings of trust, in part because it triggers oxytocin activity. When you see other people laughing, you naturally mimic them, moving muscles in your face that trigger nerves that alter your neurochemistry so that you feel happy, too; that's one reason we feel good when we are around happy people. Novelty drives up dopamine activity, making you more susceptible to romantic love. The placebo effect is real. And wet kissing transfers testosterone in saliva, helping to stimulate lust.

The black box of our humanity, the brain, is inching open. As we peer inside for the first time in human time, you and I will hold the biological codes that direct our deepest wants and feelings. We have already begun to use these codes. (I, for example, often tell people that if they want to ignite or sustain feelings of romantic love in a relationship, they should do novel and exciting things together to trigger or sustain dopamine activity.) Some one hundred million prescriptions for antidepressants are written annually in the United States, and daily many of us alter who we are in other chemical ways. As scientists learn more about the chemistry of trust, empathy, forgiveness, generosity, disgust, calm, love, belief, wanting, and myriad other complex emotions, motivations, and cognitions, even more of us will begin to use this new arsenal of weapons to manipulate ourselves and others. And as more people around the world use these hidden persuaders, one by one we may subtly change everything.

A LIVELY GAMETE MARKET

———— ‹◦› ————

HENRY HARPENDING

HENRY HARPENDING is Distinguished Professor and Thomas Chair in the Department of Anthropology, University of Utah, and the coauthor, with Gregory Cochran, of *The 10,000 Year Explosion: How Civilization Accelerated Human Evolution*.

Cheap individual genotyping will give new life to dating services and marriage arrangers. There is a market for sperm and egg donors today, but the information available to consumers about donors is limited. This industry will flourish as individual genotyping costs go down and knowledge of genomics grows.

Potential consumers will be able to evaluate not only whether a gamete provider has brown eyes, is tall or short, or has a professional degree, but also whether the donor has the appropriate MHC (major histocompatibility complex) genotypes, long or short androgen receptors, the desired dopamine receptor types, and so on. The list of criteria and the sophistication of algorithms matching consumers and donors will grow at an increasing rate in the next decade.

The idea of a "compatible couple" will have a whole new dimension. Consumers will have information about hundreds of relevant donor genetic polymorphisms to evaluate in the case of gamete markets. In marriage markets, there will be evaluation by both parties. Where will all this lead? Three possibilities come to mind:

(1) Imagine that Sally is looking at the sperm-donor market. Perhaps she is shopping for someone genetically compatible—for example, with the right MHC types. She is a homozygote for the 7R allele of the DRD4 genetic locus, so she is seeking a sperm donor homozygous for the 4R allele, so that she won't have to put up with a 7R homozygous child like she was. In other words, whether Tom or Dick is a more desirable donor depends on characteristics specific to Sally.

(2) But what if Sally values something like intelligence, which is almost completely unidimensional and of invariant polarity? Nearly everyone values high intelligence. In this case, Sally will evaluate Tom and Dick on simple scales that she shares with most other women. Tom will almost always be of higher value than Dick, and he will thus be able to obtain a higher price for his sperm.

(3) Perhaps a new president has red-headed children. Suddenly Sally, along with most other women in the market, wants red-headed children, because they are fashionable. Dick, with his red hair, is the sellout star of the sperm market but only for a short time. A cohort of children are born with red hair; then the fad goes away as green eyes, say, become the new hot seller. Dick loses his status in the market and is forced to get a real job.

These three scenarios, or any mix of them, are a possible future for love and marriage among those prosperous enough to indulge in this market. Scenario (1) corresponds to traditional views of marriage: For everyone, there is someone special and unique. Scenario (2) corresponds somewhat more closely to how marriage markets really work: Every Sally prefers rich to poor, smart to dumb, and a BMW to a Yugo. Scenario

(3) is close to one mechanism of what biologists call sexual se-lection: Male mallards have green heads essentially because it is just the fashion. I would not wager much on which of these scenarios will dominate the coming gamete market, but I favor scenario (2).

IMMORTAL COGNITION, BOUNDLESS HAPPINESS

———◇———

MARCO IACOBONI

MARCO IACOBONI is a neuroscientist at the UCLA Brain Mapping Center and the author of *Mirroring People: The New Science of How We Connect with Others.*

Life expectancy has dramatically increased over the last hundred years. At the beginning of the last century, the world average life expectancy was thirty to forty years of age, while the current average is almost seventy. Unfortunately there are still great variations in life expectancy among countries (Guess what? People in more developed countries live longer.) and within countries. (Guess what? Wealthier people live longer.) Today, people in the wealthier strata of developing countries can expect to live to more than eighty years old.

While the disparity in life expectancy is a policy issue (not discussed here), the overall dramatic increase in life expectancy brings out some interesting science issues. How can we fight the cognitive decline associated with aging, a side effect of the nice fact that we live longer? How can we fix mood disorders often associated with a general cognitive decline? The real game changer will be the immortal cognition (well, not really immortal, but close enough) and boundless happiness (okay, again, not really boundless, but close enough) provided by painless brain stimulation.

Today we have two main ways of stimulating the brain

painlessly and noninvasively: transcranial magnetic stimulation (TMS) and transcranial direct current stimulation (TDCS). TMS stimulates the brain by inducing local magnetic fields over the scalp, which in turn induce electric currents in the brain, whereas TDCS uses weak direct currents. There are also many other ways of stimulating the brain, and obviously many brain areas can be stimulated. We will be able to significantly delay cognitive decline and improve mood by stimulating those brain areas collectively called association cortices. Association cortices connect many other brain areas; their name comes from the fact that they associate many brain areas in neural networks. There are two main association cortices: those in the front of the brain, called anterior association cortices, and those in the back, called posterior association cortices. TMS has already been experimentally used for some years to treat depression by stimulating the anterior association cortices. The results are so encouraging that TMS is now an approved treatment for depression in many countries. (The Food and Drug Administration approved it for the United States in October 2008.) I believe we will see in the next two decades a great improvement in our ability to stimulate the brain to treat mood disorders. We will improve the hardware and the stimulation protocols (how frequently we stimulate and for how long). We will also improve our ability to target specific parts of the anterior association cortices using brain imaging. Each brain is slightly different, in both anatomy and physiological responses. Brain stimulation coupled with brain imaging will allow us to design specific treatments tailored to specific individuals, resulting in highly effective treatments.

The posterior association cortices—the ones in the back of the brain—are the first ones affected by Alzheimer's disease. They also show reduced activity in the less dramatic cognitive decline often associated with aging. Brain stimulation will facilitate the activity of the posterior association cortices in the elderly by inducing synchronized firing of many neurons at specific frequencies. Synchronous neuronal firing at certain frequencies is thought to be critical for perceptual and cognitive processes. Our aging brain will get its synchronized neuronal firing going, thanks to brain stimulation.

A final touch—a critical one, I would say—will be given by our ability to induce specific brain states during brain stimulation. The brain never rests, obviously; brain stimulation always stimulates the brain in a given state. The effect of brain stimulation can be thought of as the interaction between the stimulation itself and the state of the brain while it is being stimulated. Stimulating the brain while inducing specific brain states in the stimulated subject (for instance, by playing games that require the subject to associate words, or showing the subject stimuli usually associated with happiness) will result in much more effective treatments of cognitive decline and mood disorder.

This will be a real game changer. If my prediction is correct, we will also see dramatic changes in policy. People won't tolerate being excluded from the beneficial effects of brain stimulation. Right now, people don't easily grasp the insidious environmental factors or subtle differences in health care that result in dramatic individual differences in the long term (approximately ten years of life between the wealthy and the poor living in the same country), but they will immediately grasp the beneficial effects of brain stimulation and will demand not to be excluded anymore. That's also a game changer.

A FAREWELL TO HARM

———◁◦▷———

KARL SABBAGH

KARL SABBAGH is a writer and television producer and the author of *The Riemann Hypothesis: The Greatest Unsolved Problem in Mathematics*.

Much of the misery in the world today—as it always has been—is due to the human propensity for contemplating, or actually committing, violence against another human being. It's not just assaults and murders that display that propensity. Someone who designs a weapon, punishes a child, declares war, or leaves a hit-and-run victim by the side of the road has defined "harming another human being" as a justifiable action for himself. How different the world would be if, as a biologically determined characteristic of future human beings, there were such a cognitive inhibition to such actions that people would be incapable of carrying them out, just as most of us are incapable of moving our ears.

It must be the case that in the brains of everyone, from abusive parents and rapists to arms dealers and heads of state, there can arise a concatenation of nerve impulses which allows someone to see as "normal"—or at least acceptable—the mutilation, maiming, or death of another for one's own pleasure or benefit. Suppose the pattern of that series of impulses was analyzable *exactly*, with future developments of functional magnetic resonance imaging (fMRI), PET scans, or technology as yet uninvented. Perhaps every decision to kill or harm another person can be traced to a series of nerve impulses that arise in brain center A, travel in a microsecond to areas B, C, and D, inhibit areas E and F, and lead to a previously unacceptable decision becoming acceptable. Perhaps we would discover a common factor between the brain patterns of someone who is about to murder a child and that of a head of state signing a bill to initiate a nuclear weapons program or an engineer designing a new type of cluster bomb. All of them recognize at some intellectual level that it is perfectly all right for their actions to cause harm or death to another human. The brains of all of them, perhaps, experience pattern D, the "death pattern."

If such a specific pattern of brain activity were detectable, could methods then be devised that prevented or disrupted it whenever it was about to arise? The most plausible—and least socially acceptable—of these methods would entail everyone wearing microcircuit-based devices that detected the pattern and suppressed or disrupted it, such that anyone in whom the impulse arose would instantaneously lose any will to carry it out. Less plausible, but still imaginable, would be some sophisticated chemical suppressant of pattern D, genetically engineered to act at specific synapses or on specific neurotransmitters and delivered in some way that reached every single member of the world's population. The pattern D suppressant could be deployed as a water additive like chlorine (acceptable now to prevent deaths from dirty water), or as inhalants sprayed from the air, or in genetically modified foodstuffs—even, perhaps, as an alteration of the germ-cell line in one generation that would forever remove pattern D from future generations.

Rapes would be defused before they happened; soldiers, if there were still armies, would be inhibited from firing as their trigger fingers tightened—except there would be no one to fire at, since enemy soldiers, insurgents, or terrorists would themselves be unable to carry their violent acts to completion.

Would the total elimination of murderous impulses from the human race have a downside? Well, of course, one single person who escaped the elimination process could then rule the world. He—let's say it's a man—could oppress and kill with impunity, since no one else would have the will to kill him. Measures would have to be devised to deal with such a situation. Such a person would be so harmful to the human race that, perhaps, plans would have to be laid to control him if he should arise. Tricky, this one, since he couldn't be killed outright: There would be no one able to kill him, or even to design a machine that would kill him, as that would involve an ability to contemplate the death of another human being.

But setting that possibility aside, what would be the disadvantages of a world in which, chemically or electronically, the ability to kill or harm another human being would be removed from all people? Surely only good could come from it. Crimes motivated by greed would still be possible, but robberies would be achieved with trickery rather than at the point of a pistol; gang members might attack one another with insults and taunts rather than razors or coshes; governments might play chess to decide contentious border issues; and deaths from road accidents would go down, because even the slightest thought about one's own behavior causing the death of another would be so reminiscent of pattern D that we would all drive much more carefully to avoid it. Deaths from natural disasters would continue, but charitable giving and international aid in such situations would soar, as people realized that not helping to prevent them in future would be almost as bad as the old and now-defunct habit of killing people.

A method to eliminate pattern D will lead to the most significant change ever in the way humans—and therefore societies—behave. And somewhere today, in the fields of neurobiology or genetic modification, the germ of that change may already be present.

GOD NEED NOT ACTUALLY EXIST TO HAVE EVOLVED

—◇—

JESSE BERING

JESSE BERING is a psychologist, director of the Institute of Cognition and Culture at the Queen's University, Belfast, and a columnist for *Scientific American* ("Bering in Mind").

What if I were to tell you that God was all in your mind? That God, like a tiny speck floating at the edge of your cornea producing the image of a hazy, out-of-reach orb accompanying your every turn, was in fact an illusion, a psychological blemish etched onto the core cognitive substrate of your brain? It may feel like there is something grander out there—watching, knowing, caring. Perhaps even judging. But in fact there is only the air you breathe. Consider, briefly, the implications of seeing God this way, as a sort of scratch on our psychological lenses rather than the enigmatic figure out there in the heavenly world most people believe him to be. Subjectively, God would still be present in our lives. In fact, rather annoyingly so. As a way of perceiving, he would continue to suffuse our experiences with an elusive meaning and give us the sense that the universe is communicating with us in various ways. But objectively, the notion of God as an illusion is a radical and, some would say, even dangerous idea, since it raises important questions about God as an autonomous, independent agent that lives outside human brain cells.

In fact, the illusion of God is more plausible a notion than some related thought experiments, such as the possibility that our brains are sitting in an electrified vat somewhere and we're merely living out simulated lives. In contrast to the vat exercise, or some other analogy to the science-fiction movie *The Matrix*, it is uncontroversial to say that our species' ability to think about God (even an absent God) is made possible only by our naturally derived brains—in particular, by virtue of the fact that our brains have evolved over the eons in the unusual manner they have. In philosophical discourse, the idea that God is an illusion would be a scientifically inspired twist on a very ancient debate, since it deals with the nature and veridicality of God's actual being.

That's all very well, you may be thinking, but perhaps God isn't an illusion at all. Rather than a scratch on our psychological lenses, our brain's ability to reason about the supernatural—about such things as purpose, the afterlife, destiny—is God's personal signature on our brains. One can never rule out the possibility that God microengineered the evolution of the human brain so that we have come to see him more clearly—a sort of divine LASIK procedure, or a scraping off of the bestial glare that clouds the minds of other animals. Some scholars, such as the psychologists Justin Barrett and Michael Murray, hold something like this "theistic evolution" view. Yet as a psychological scientist who studies religion, I take explanatory parsimony seriously. After all, parsimony is the basic premise of Ockham's razor, the cornerstone of scientific enquiry. Ockham's razor holds that of two equally plausible theories, science shaves off the extra fat by favoring the one that makes the fewest unnecessary assumptions. And in the natural sciences, the concept of God as a causal force tends to be an unpalatable lump of gristle. Although treating God as an illusion may not be entirely philosophically warranted, therefore, it is in fact a scientifically valid treatment. Because the human brain (like any physical organ) is a product of evolution, and because natural selection works without recourse to intelligent forethought, this mental apparatus of ours evolved to think about God quite without need of his consultation, let alone his being real.

Indeed, the human brain has many such odd quirks that systematically alter, obscure, or misrepresent entirely the world outside our heads. That's not a bad thing necessarily, nor does it imply poor adaptive design. You have undoubtedly seen your share of optical illusions before, such as the famous Müller-Lyer image wherein a set of arrows of equal length with their tails pointing in opposite directions creates the subjective impression that one line is longer than the other. You know that the lines are of equal length, yet despite this knowledge your mind does not allow you to perceive the image this way. There are also well-documented social-cognitive illusions that you may not be as familiar with. For example, David Bjorklund, a developmental psychologist, reasons that young children's overconfidence in their own abilities keeps them engaging in challenging tasks rather than simply giving up when they fail. Ultimately, with practice and over time, children's actual skills can ironically begin to more closely approximate those earlier, favorably warped self-judgments. Similarly, evolutionary psychologists David Buss and Martie Haselton argue that men's tendency to overinterpret women's smiles as sexual overtures prompts them to pursue courtship tactics more often, sometimes leading to real reproductive opportunities with friendly women.

In other words, whether our beliefs about the world "out there" are accurate matters little; rather, psychologically speaking, it's whether they work for us—or for our genes—that counts. As you read this, cognitive scientists are inching their way toward a more complete understanding of the human mind as a reality-ending prism. What will change everything? The looming consensus, among those who take Ockham's razor seriously, that the existence of God is a question for psychologists and not physicists.

PROOF OF THE RIEMANN HYPOTHESIS

---◦▸---

CLIFFORD A. PICKOVER

CLIFFORD A. PICKOVER is a science writer, an editor at the IBM T. J. Watson Research Center, and the author of *The Math Book*.

Many mathematical surveys indicate that the proof of the Riemann hypothesis is the most important open question in mathematics. The rapid pace of mathematics, along with computer-assisted mathematical proofs and visualizations, leads me to believe that this question will be resolved in my lifetime. Math aficionado John Fry once said that he thought we would have a better chance of finding life on Mars than finding a counterexample for the Riemann hypothesis.

In the early 1900s, British mathematician Godfrey Harold Hardy sometimes took out a quirky form of life insurance when embarking on ocean voyages. In particular, he would mail a postcard to a colleague on which he would claim to have found the solution of the Riemann hypothesis. Hardy was never on good terms with God and felt that God would not let him die in a sinking ship while Hardy was in such a revered state, with the world always wondering if he had really solved the famous problem.

The proof of the Riemann hypothesis involves the zeta function, which can be represented by a complicated-looking curve that is useful in number theory for investigating properties of prime numbers. Written as $f(x)$, the function was originally defined as the infinite sum:

$$f(x) = 1 + (1/2)^x + (1/3)^x + (1/4)^x + \ldots \text{ and so on.}$$

When $x = 1$, this series has no finite sum. For values of x larger than 1, the series adds up to a finite number. If x is less than 1, the sum is again infinite. The complete zeta function, studied and discussed in the literature, is a more complicated function that is equivalent to this series for values of x greater than 1, but it has finite values for any real or complex number, except for when the real part is equal to 1. We know that the function equals zero when x is -2, -4, -6, and so on. We also know that the function has an infinite number of zero values for the set of complex numbers, the real part of which is between zero and 1—but we do not know exactly for what complex numbers these zeros occur. In 1859, mathematician Georg Bernhard Riemann conjectured that these zeros occur for those complex numbers the real part of which equals 1/2. Although vast numerical evidence exists that favors this conjecture, it is still unproved.

The proof of Riemann's hypothesis will have profound consequences for the theory of prime numbers and for our understanding of the properties of complex numbers. A generalized version of the hypothesis, when proved true, will allow mathematicians to solve numerous important mathematical problems. Amazingly, physicists may have found a mysterious connection between quantum physics and number theory through investigations of the Riemann hypothesis. I do not know if God is a mathematician, but mathematics is the loom upon which God weaves the fabric of the universe.

Today, over eleven thousand volunteers around the world are working on the Riemann hypothesis, using a distributed computer software package at Zetagrid.Net to search for the zeros of the Riemann zeta function. More than one billion zeros for the zeta function are calculated every day.

In modern times, mathematics has permeated every field of scientific endeavor and plays an invaluable role in biology, physics, chemistry, economics, sociology, and engineering. Mathematics can be used to help explain the colors of a sunset or the architecture of our brains. Mathematics helps us build supersonic aircraft and roller coasters, simulate the flow of Earth's natural resources, explore subatomic quantum realities, and image faraway galaxies. Mathematics has changed the way we look at the cosmos.

Physicist Paul Dirac once noted that the abstract mathematics we study now gives us a glimpse of physics in the future—and in fact, his equations predicted the existence of antimatter, which was subsequently discovered. Similarly, mathematician Nikolai Lobachevsky said, "There is no branch of mathematics, however abstract, which may not someday be applied to the phenomena of the real world."

THE REALITY OF TIME

———◦———

LEE SMOLIN

LEE SMOLIN is a physicist at the Perimeter Institute for Theoretical Physics in Waterloo, Ontario, and the author of *The Trouble with Physics: the Rise of String Theory, the Fall of a Science, and What Comes Next.*

I would like to describe a change in viewpoint that I believe will alter how we think about everything from the most abstract questions on the nature of truth to the most concrete questions in our daily lives. This change comes from the deepest and most difficult problems facing contemporary science: those having to do with the nature of time.*

The problem of time confronts us at every key juncture in fundamental physics: What was the Big Bang and could something have come before it? What is the nature of quantum physics and how does it unify with relativity theory? Why are the laws of physics we observe the true laws, rather than other possible laws? Might the laws have evolved from different laws in the past?

After a lot of discussion and argument, it is becoming clear to me that these key questions in fundamental physics come down to a very simple choice, having to do with the answers to two simple questions: What is real? And what is true?

Many philosophies and religions offer answers to these questions, and most give the same answer: Reality and truth transcend time. If something is real, it has a reality that continues forever, and if something is true, it is not just true now, it has always been true and always will be. The experience we have of the world existing within a flow of time is, according to some religions and many contemporary physicists and philosophers, an illusion. Behind that illusion is a timeless reality: in modern parlance, the block universe. Another manifestation of this ancient view is the currently popular idea that time is an emergent quality, not present in the fundamental formulation of physics.

The new viewpoint is the direct opposite. It asserts that what is real is only what is real in the moment, which is one of a succession of moments. It is the same for truth: What is true is only what is true in the moment. There are no transcendent, timeless truths.

There is also no past. The past lives only as part of the present, to the extent that it gives us evidence of past events. And the future is not yet real, which means that it is open and full of possibilities, only a small set of which will be realized. Nor, in this view, is there any possibility of other universes. All that exists must be part of this universe that we find ourselves in, at this moment.

This view changes everything, beginning with how we think of mathematics. In this view, there can be no timeless Platonic realm of mathematical objects. The truths of mathematics, once discovered, are certainly objective. But mathematical systems have to be invented—or evoked—by us. Once brought into being, there are an infinite number of facts that are true about mathematical objects: facts that further investigation might discover. There are an infinite number of possible axiomatic systems we might so evoke and explore, but the fact that different people will agree on what has been shown about them does not imply that they existed before we evoked them.

I used to think that the goal of physics was the discovery of a timeless mathematical equation that was isomorphic to the history of the universe. But if there is no Platonic realm of timeless mathematical objects, this is just a fantasy. Science is, then, only about what we can discover is true in the one real universe we find ourselves in.

More specifically, this view challenges how we think about cosmology. It opens up new ways to approach the deepest questions, such as why the laws we observe are true and not others, and what determined the initial conditions of the universe. The philosopher Charles Sanders Peirce wrote in 1893 that the only way of accounting for which laws were true would be through a mechanics of evolution, and I believe this remains true today. But the evolution of laws requires time to be real. Furthermore, there is, I believe, evidence on technical grounds that the correct formulations of quantum gravity and cosmology will require the postulate that time is real and fundamental.

But the implications of this view will be far broader. For example, in neoclassical economic theory, which is anchored in the study of equilibria of markets and games, time is largely abstracted away. The fundamental results on equilibria by Kenneth Arrow and Gerard Debreu assume that there are fixed and specifiable lists of goods and strategies and that each consumer's tastes and preferences are unchanging.

But can this be completely correct, if growth is driven by opportunities that suddenly appear from unpredictable discoveries of new products, new strategies, and new modes of organization? Getting economic theory right has implications for a wide range of policy decisions, and how time is treated is a key issue. An economics that assumes that we cannot predict key innovations must be very different from one that assumes all is knowable at any time.

The view that time is real and truth is situated within the moment further implies that there is no timeless arbiter of meaning and no transcendent or absolute source of values or ethics. Meaning, values, and ethics are all things that we humans project into the world. Without us, they don't exist.

This means that we have tremendous responsibilities. Both mathematics and society are highly constrained, but within those constraints there are an infinitude of possibilities, only a few of which can be evoked and explored in the finite time we have. Because time is real and the future does not yet exist, the imaginative and social worlds in which we will live are to be brought into being by the choices we will make.

The views expressed here are the result of a collaborative project with the Brazilian philosopher Roberto Mangabeira Unger.

THE EXISTENCE OF ADDITIONAL SPACETIME DIMENSIONS

◄○►

GINO SEGRÈ

GINO SEGRÈ is a physicist at the University of Pennsylvania and the author of *Faust in Copenhagen: A Struggle for the Soul of Physics*.

Einstein's general theory of relativity, first presented in the fall of 1915, and his earlier special theory of relativity have changed very little of our day-to-day world, but they have radically altered the way we think about both space and time and have also launched the modern theory of cosmology. If, in the near future, we discover additional spacetime dimensions, we will undergo a shift in our perceptions every bit as radical as the one experienced almost a hundred years ago. Though proof of their existence would necessarily alter our view of the universe, there is also a way in which our psyches would be changed. I believe we would gain a new confidence that great, almost unimaginable, phenomena are yet to be discovered. We would also realize once again the power that lies in a few simple equations, in the tools we can build to test them, and in the human imagination.

At the November 6, 1919, joint meeting of the Royal Society and the Royal Astronomical Society, Sir Frank Watson Dyson reported on the observations of starlight made during the previous May's solar eclipse. "After a careful study of the plates, I am prepared to say that they confirm Einstein's prediction. A very definite result had been obtained, that light is deflected in accordance with Einstein's law of gravitation." Sir Joseph John Thomson, presiding, afterward called the result "one of the highest achievements of human thought." It was a triumphant moment for both theoretical physics and observational astronomy.

A few years after that momentous meeting, a German mathematician and a Swedish physicist, Theodor Kaluza and Oskar Klein, reached a striking conclusion. They noticed that the equations of general relativity, when solved in five rather than four dimensions, led to additional solutions that were identical to the well-known Maxwell equations of electromagnetism. Since the apparent fifth dimension had not—and still has not—been observed, a necessary postulate for this theory to correspond to possible reality was that the fifth dimension was curled up so tightly that any motion in its direction had not been detected.

Einstein, finding this extension of his general theory of relativity extraordinarily attractive, tried more than once—without success—to make it part of his lifelong dream of a unified field theory of interactions. But this avenue of research fell into some disfavor in the decades immediately after World War II, during which theoretical physics turned its attention to other matters. It returned with a vengeance in the late 1970s, gaining momentum in the 1980s as physicists began to seriously examine theories that could unite all fundamental interactions into one comprehensive scheme. The rising popularity of superstring theory, mathematically consistent only if additional spacetime dimensions are present, has provided the decisive impetus for such considerations.

There are striking differences from the 1915 situation, most particularly the lack of a clear test for the detection of extra dimensions. The novel theories now in fashion do predict that additional particles must be present in nature because of these extensions of space and time, but since the mass of these particles is related to the unknown scale of the extra dimensions, it too remains unknown. Roughly speaking, the smaller the one, the larger the other. Nevertheless, the hunt has begun. We are beginning to see in the literature publications from major laboratories with titles such as "Search for Gamma Rays from the Lightest Kaluza-Klein Particle"—that being the name frequently given to the as-yet-undiscovered particles associated with extra dimensions.

These searches are largely motivated by the desire to identify dark matter, which is estimated to be several times more plentiful in our universe's makeup than all known species of matter. Kaluza-Klein particles are one possible candidate, perhaps hard to distinguish from other candidates even if found. Challenges abound, but the stakes are high.

BLACK HOLES: THE ULTIMATE GAME CHANGER?

PAUL J. STEINHARDT

PAUL J. STEINHARDT is the Albert Einstein Professor of Science at Princeton University and coauthor, with Neil Turok, of *Endless Universe: Beyond the Big Bang*.

One of the sacred principles of physics is that information is never lost. It can be scrambled, encrypted, dissipated, and shredded, but never lost. This tenet underlies the second law of thermodynamics and a concept called unitarity, an essential component of unified theories of particles and forces. Discovering a counterexample or new ways to preserve information could be a real game changer—one that alters our understanding of the fundamental laws of nature, transforms our concept of space and time, triggers a reconstruction of the history of the universe, and leads to new prognostications about its future.

There is a real chance of breakthrough in the foreseeable future, as theorists converge on one of the greatest threats to information preservation: black holes. According to Einstein's general theory of relativity, a black hole forms when matter is so concentrated that nothing, not even light, can escape its gravitational field. Any information that passes through the event horizon surrounding the black hole—the "point of no return"—is lost forever to the outside world. Suppose, for example, that Bob pilots a spaceship into the black hole carrying along three books of his choice. It appears that the contents of the three books vanish. Either that or Einstein's general theory of relativity is wrong.

There is nothing shocking about having to correct Einstein's general theory of relativity. It's known to be missing an essential element: quantum physics. Einstein, and generations of theorists since, have sought an improved theory of gravity: one that incorporates quantum physics in a way that is mathematically and physically consistent. String theory and loop quantum gravity are the most recent attempts.

There is no doubt that quantum physics alters the event horizon and the evolution of a black hole in a fundamental way, as was first pointed out in the work of Jacob Bekenstein, Gary Gibbons, and Stephen Hawking in the 1970s. According to quantum physics, matter and energy are composed of discrete chunks known as quanta (such as electrons, quarks, and photons) whose positions and velocities undergo constant random fluctuations. Even empty space—a pure vacuum—is full of microscopic fluctuations that create and annihilate quanta/antiquanta pairs. The seething vacuum just outside the event horizon occasionally produces such a pair—an electron-positron duo, say—in which one escapes and one falls into the black hole. From afar, it appears that the black hole has radiated a particle. This phenomenon continually repeats, producing a spectrum of particles known as Hawking radiation, whose properties are similar to the thermal radiation emitted by a hot body. Very slowly, the black hole radiates away energy and shrinks in mass and size until—well, here is where the story really begins to get interesting.

Thermal radiation depends only on the temperature of the emitting body and provides no other details about the body itself. So if Hawking radiation is truly thermal, then the information inside the black hole is truly lost. For the last decade, though, leading physicists, including Hawking, Gerard 't Hooft, and Leonard Susskind, fiercely debated (and even bet on) the outcome—Susskind refers to the debate as the Black Hole War. Aided by new theoretical tools developed by Juan Maldacena and other string theorists, physicists discovered that Hawking radiation is not quite thermal after all. The radiation deviates by a tiny amount from a perfectly thermal signal, and the tiny deviation incorporates information about whatever was inside. The contents of Bob's three books, for example, are not lost forever, although the information dribbles out incredibly slowly and is unimaginably scrambled. Thus victory was declared in the Black Hole War.

But it may be an uneasy peace, for there remains the question of what happens to information after it falls into the horizon. This is a reasonable question because, curiously enough, passage through the horizon can be unremarkable if the black hole is very big. There are no signposts indicating to Bob that he has passed the point of no return, and his books remain intact. Now suppose Bob scribbles some notes in the margins of his books. What happens to this information?

Here there is a diversity of views. Some suggest that this information, too, is radiated away through the Hawking process and the black hole simply disappears. Some suggest that quantum physics makes the event horizon penetrable, so that some information is radiated by the Hawking process but some escapes directly. Yet others suggest that the information is copied; one copy is radiated away and the other strikes the singularity, entering a new section of spacetime that is causally disconnected from observers outside the black hole, so the two copies never meet.

Theorists have recently developed a number of new theoretical tools to attack the problem and are hard at work. Although the subject lies in the domain of quantum gravity, the implications for other fields—including my own, cosmology—will be profound. The answer will shape any future formulation of the laws of thermodynamics, quantum gravity, and unified field theory. Since scrambling information—aka entropy—determines the direction of the arrow of time, the results may inform us how time first emerged at the cosmic singularity known as the Big Bang. Or, if it proves possible for copies to bounce from the black hole singularity to a separate piece of spacetime, this could totally change our view of the Big Bang. In particular, it would lend support to recent ideas suggesting that the Big Bang was not the beginning of space and time, that the large-scale properties of the universe were shaped by events before the Big Bang, and that these conditions (a form of information) were transmitted across the cosmic singularity into a new phase of expansion. In fact, if information is forever preserved across singularities, the universe may undergo regularly repeating cycles of Big Bangs, expansion, and Big Crunches, forever into the past and into the future. To me, a breakthrough with these kinds of implications would be the ultimate game changer.

BETTER MEASUREMENTS

—◦—

GREGORY COCHRAN

GREGORY COCHRAN is a consultant with Adaptive Optics, an adjunct professor of anthropology at the University of Utah, and the coauthor, with Henry Harpending, of *The 10,000 Year Explosion: How Civilization Accelerated Human Evolution.*

Our most reliable engine of change has been increased understanding of the physical world. First it was Galilean dynamics and Newtonian gravity, then electromagnetism, later quantum mechanics and relativity. In each case, new observations revealed new physics—physics that went beyond the standard models, physics that led to new technologies and new ways of looking at the universe. Often those advances were the result of new measurement techniques. The Greeks never found artificial ways of extending their senses, which hobbled their protoscience. But ever since Tycho Brahe, a man with a nose for instrumentation, better measurements have played a key role in Western science.

We can expect significantly improved observations in many areas over the next decade. Some of that is due to sophisticated, expensive, and downright awesome new machines. The Large Hadron Collider should begin producing data this year and maybe even information. We can scan the heavens for the results of natural experiments you wouldn't want to try in your backyard—events that shatter suns and devour galaxies—and we're getting better at that. That means devices like the 30-meter telescope under development by a Caltech-led consortium, or the 100-meter OWL (Overwhelmingly Large Telescope) under consideration by the European Southern Observatory. Those telescopes will actively correct for the atmospheric fluctuations that make stars twinkle—but that's almost mundane, considering that we have a neutrino telescope at the bottom of the Mediterranean and another buried deep in Antarctic ice. We have the world's first real gravitational observatory (LIGO, the Laser Interferometer Gravitational-Wave Observatory) running now, and planned improvements should increase its sensitivity enough to study cosmic fender-benders in the neighborhood, as (for example) when two black holes collide.

There's no iron rule ensuring that revolutionary discoveries must cost an arm and a leg. Ingenious experimentalists are testing quantum mechanics and gravity in tabletop experiments as well. They'll find surprises. When you think about it, even historians and archaeologists have a chance of shaking gold out of the physics tree: We know the exact date of the Crab Nebula supernova from old Chinese records, and with a little luck we'll find some cuneiform tablets that give us some other astrophysical clue, as well as the real story about the Battle of Kadesh.

We have a lot of all-too-theoretical physics under way, but there's a widespread suspicion that the key shortage is data, not mathematics. The universe may not be stranger than we can imagine, but it's entirely possible that it's stranger than we have imagined thus far. We have string theory, but what Bikini test has it brought us? Experiments led the way in the past, and they will lead the way again.

We will probably discover new physics in the next generation, and there's a good chance that the world will, as a consequence, become unimaginably different. For better or worse.

WE ARE LEARNING TO MAKE PHENOTYPES

———◇———

MARK PAGEL

MARK PAGEL is an evolutionary biologist at Reading University, UK, and an external professor at the Santa Fe Institute.

We all develop from a single cell known as a zygote. This zygote divides and becomes two cells, then four, eight, and so on. At first, most of the cells are alike, but as this division goes on, something wondrous occurs: The cells begin to commit themselves to different fates—as eyes or ears, or livers or kidneys, or brain or blood cells. Eventually they produce a body of unimaginable complexity, making things like supercomputers and space shuttles look like Lego toys. No one knows how they do it. No one is there to tell the cells how to behave; there is no homunculus directing cellular traffic, and no template to work from. It just happens.

If scientists could figure out how cells enact this miracle of development, they could produce phenotypes (the outward form of our bodies) at will and from scratch, or at least from a zygote. This, or something close to it, will happen in our lifetimes. When we perfect this developmental process—and we are well on the way—we will be able to re-create ourselves, even redefine the nature of our lives.

The problem is that development isn't just a matter of finding a cell and getting it to grow and divide. As our cells differentiate into our various body parts, they lose what is known as their potency—that is, they "forget" how to go back to their earlier states, in which all fates were possible. When we cut ourselves, the skin nearby knows how to grow back, erasing all or most of the damage. But this happens only on a very local scale: If you cut off your arm, it does not grow back. What scientists are learning, bit by bit, is how to reverse cells to their earlier potent states and reprogram them so that they *will* replace a limb.

Every year brings new discoveries and new successes. Cloning is one of the more visible. At the moment, most cloning is a bit of a cheat, achieved by taking special cells from an adult animal's body which still retain some of their potency. But this will change as cell reprogramming becomes possible, and the consequences could be alarming. Someone might be able to clone you by collecting a bit of your hair or other cells left behind when you touch something. Why someone would want to do this—and wait for you to grow up—might limit the practice, but it could happen. You could become your own "father," or at least a very grown-up twin.

More in the realm of the everyday, and of real consequence, is that once we can reprogram cells, whole areas of science and medicine—including aging, injury, and disease—will vanish or become unimportant. The contentious debate about the morality of work on embryonic stem cells arose solely because scientists want a source of "totipotent" cells—cells that haven't committed themselves. Embryos are full of them. Once scientists can return cells to their totipotent state—or even to a "multipotent" state (in which a cell is not quite fully committed)—stem-cell research will become unnecessary. This could happen within a decade.

Schoolchildren know that some lizards and crabs can regrow limbs. What they are not taught is that this is because the cells of those species retain some or all of their potency. Ours don't, which makes car crashes, ski accidents, gunshot wounds, and growing old a nuisance. But once we unlock the door of development, we will be able to regrow our limbs, heal our wounds, and much more. The limbs (and organs, nerves, body parts, and so on) that we regrow will be real, making those bionic parts that Anakin Skywalker gets fitted with after a light saber accident seem primitive. Organ transplants will be obsolete or just temporary; heart disease will be cured by growing a new heart. Nerve damage and paralysis will be reversible; certain brain diseases will become treatable. Some of this is already happening.

As if these developments were not life-changing enough, they will in the long run usher in an era in which our minds— what we think of as "us"—can be separated from our bodies, or nearly separated. I don't suggest that we will be able to transplant our minds to other bodies, but we will be able to introduce more and more new body parts into an existing body with a resident mind—so that we will become essentially immortal, like ancient buildings that exist only because, over the centuries, each of their many stones has been replaced.

Another intriguing aspect of reprogramming cells is that they can be induced to forget how old they are. Aging will thus become a thing of the past, if you can afford enough new pieces. We will then discover the extent to which our minds arise from perceptions of our bodies and the passage of time. If you give an old person the body of a teenager, say, will the elderly recipient start to behave and think like one? Who knows—but it will be game changing to find out.

THE NEXT STEP IN HUMAN HEALTH CARE?

———◇———

IAN WILMUT

IAN WILMUT is an embryologist, the chair of reproductive biology and director of the Scottish Centre for Regenerative Medicine at the University of Edinburgh, and the coauthor, with Roger Highfield, of *After Dolly: The Promise and Perils of Cloning.*

In 2009, we are still comparatively near the beginning of an era in which biomedical research is revolutionizing our understanding of inherited human diseases and providing the first effective treatment for some of them. This new knowledge will offer benefits at least as great as those from past biomedical research, which has significantly reduced the devastating effects of many infectious diseases. The powerful new tools that will bring this about are those used in molecular-genetic analysis and stem-cell biology.

Human health and life span in the more fortunate parts of the world have improved dramatically in the past thousand years, but in the main this is because we became better at meeting the everyday needs for survival. Over this period, humans became more effective at collecting or producing adequate supplies of food. On this timescale, it was only recently that communities recognized the need for clean water and effective sanitation to prevent infection. More recently still, methods have been developed for immunization against potential infection, and compounds have been identified as powerful antibiotics. While the authors of these essays, and the vast majority of those who read them, can take these benefits for granted, it is a sad commentary on us all that this is not true for many millions in the less fortunate parts of the world—but that is another matter.

The coming together of emerging techniques in cell and molecular biology will change our entire approach to those human diseases that are inherited rather than acquired from an infective agent such as a virus or bacterium. Inherited diseases are those that run in families because of errors in the DNA sequence of some family members. For simplicity's sake, this essay will concentrate on diseases inherited through chromosomal DNA, while acknowledging that mitochondrial DNA is also error prone and the cause of other inherited diseases.

Whereas the proportion of diseases for which the precise genetic cause has been identified is increasing because of the power of modern molecular analysis, it is still small. Even more important, how the genetic error causes the disease symptoms is known in very few cases. This has been a major limiting factor in the development of effective treatments, because the goal of present treatments is not to correct the error in the DNA but to prevent the development of the symptoms.

One advantage of the new tools is that one need not identify the genetic error to be able to identify compounds that can prevent the development of symptoms. This opportunity arises from the revolutionary new technique by which stem cells—cells able to form all types of body tissues—are formed from cells taken from adults. Shinya Yamanaka, of Kyoto University, was the first to show that a simple procedure could achieve this extraordinary change, and he named these cells "induced pluripotent cells," because of their ability to form all tissues.

Many laboratories are now using induced pluripotent stem cells to study inherited diseases such as the neurodegenerative disease ALS (amyotrophic lateral sclerosis). Pluripotent cells from ALS patients are turned into the various neural populations affected by the disease and contrasted with the same cell population from healthy donors. Discovery of the molecular cause of the diseases will involve analyses of gene function in the diseased cells in many ways. There is then the practical issue of devising a test to discover if potential drugs can prevent the development of symptoms. This can be used as the basis of tests that can be carried out by robots able to screen thousands of compounds every week. Many further studies will then be required before any new medicine can be used to treat patients.

It is also likely that these therapies will be effective in treating related cases for which there is no evidence of a genetic cause. In the case of ALS, it is estimated that fewer than 10 percent of cases are inherited. ALS should be considered as a family of diseases, because it reflects errors in several different genes. Recent studies have revealed an unusual distribution of a particular protein within the cells of many patients. This occurred in inherited cases associated with all except one of these genes and, in addition, occurred in several patients in whom there was no evidence of an inherited effect. While this pattern may not occur with all inherited diseases, the observation encourages the hope that treatments developed through research with inherited cases will often be equally effective for cases in which there is no genetic effect.

While I have separated infectious and inherited diseases, in reality there is a considerable overlap. New understanding of the molecular and cellular mechanisms that govern normal development and health will also provide the basis for novel treatments for infectious diseases. For example, understanding the development and function of the immune system may reveal new approaches to the treatment of diseases such as HIV-AIDS.

I find it exciting indeed to think that in my lifetime, effective treatments will be available for some of the many hundreds of inherited diseases. Their devastating effect on patients and their families will be greatly reduced, or even removed, just as earlier research banished infections such as polio, tuberculosis, and such childhood diseases as measles and mumps.

BROADENING THE SPECTRUM OF INFECTIOUS CAUSATION

PAUL EWALD

PAUL EWALD is a professor of biology at Amherst College and the author of *Plague Time: The New Germ Theory of Disease*.

Thirteen decades ago, Louis Pasteur and Robert Koch led an intellectual revolution referred to as the germ theory of disease, which proposed that many common ailments were caused by microbes. Since then, the accepted spectrum of infectious causation has been increasing steadily and dramatically. The diseases that are most obviously caused by infection were accepted as such by the end of the nineteenth century; almost all of them were acute diseases. Acute diseases with a transmission twist—mosquito-borne malaria, for example—were accepted a bit later, at the beginning of the twentieth century. Since the early twentieth century, the spectrum has broadened, mostly by recognition of the infectious causation of chronic diseases. The first of these had distinctly infectious acute phases, which made infectious causation of the chronic disease more obvious; infectious causation of shingles, for example, was made more apparent by its association with chicken pox. Over the past thirty years, the spectrum of infectious causation has broadened mostly through inclusion of chronic diseases *without* obvious acute phases. With years or even decades between the onset of infection and the onset of such diseases, demonstration of infectious causation is difficult.

Technological advances have been critical to resolving the ambiguities associated with such cryptic infectious causation. In the early 1990s, Kaposi's sarcoma–associated herpes virus was discovered by using a molecular technique that stripped away the human genetic material from Kaposi's sarcoma cells and found what remained. A similar approach revealed hepatitis C virus in blood transfusions. In these cases, there were strong epidemiological signs that an infectious agent was present. When the cause was discovered, acceptance did not have to confront the barrier of entrenched opinions favoring noninfectious causes. If such special interests are present, the evidence has to be proportionately more compelling: Such is the case for schizophrenia, atherosclerosis, Alzheimer's disease, breast cancer, and many other chronic diseases that are now the focus of vehement disagreements.

Advances in molecular/bioinformatic technology are poised to help resolve these controversies. This potential is illustrated by two discoveries that seem cutting-edge now but will soon be considered primitive first steps. About a decade ago, one member of a team at Stanford University scraped spots on two teeth of another team member and amplified the DNA from the scrapings. The team found sequences that were sufficiently unique to represent more than thirty new species. This finding hinted at the magnitude of the challenge—tens of thousands, or perhaps even hundreds of thousands, of viruses and bacteria may need to be considered to evaluate hypotheses of infectious causation.

The second discovery provides a glimpse of how this challenge may be addressed. Samples from prostate tumors were tested on a microarray containing twenty thousand DNA snippets from all known viruses. The results documented a significant association with an obscure retrovirus related to one that normally infects mice. If this virus is a cause of prostate cancer, it causes only a small portion occurring in men with a particular genetic background. Other viruses have been associated with prostate cancer in patients without this genetic background. So, not only may thousands of viruses need to be tested to find one correlated with a chronic disease, but even then it may be one of perhaps many different infectious causes.

The problems of multiple pathogens and ingrained predispositions are now coming to a head in research on breast cancer. Currently, three viruses have been associated with breast cancer: mouse mammary tumor virus, Epstein-Barr virus, and human papillomavirus. Researchers are still arguing about whether these correlations reflect causation. If they do, these viruses account for somewhere between half and about 95 percent of incidences of breast cancer, depending on the extent to which they act synergistically. Undoubtedly, array technology will soon be used to assess this possibility and to identify other viruses that may be associated with breast cancers.

There is a caveat. These technological advances provide sophisticated approaches to identifying correlations between pathogens and disease. They do not bridge the gulf between correlation and causation. One might hope that with enough research, all aspects of the pathological process will be understood, from the molecular level up to the whole patient. But as one moves from the molecular to the macro levels, the precision of interpretation becomes confounded by the complex web of interactions that intervene, especially in chronic diseases. Animal models are generally inadequate for studying chronic human diseases, because the disease in animals is almost never quite the same as the human disease. The only way out of this conundrum, I think, will be to complement the technological advances in identifying candidate pathogens with clever clinical trials. These clinical trials will need to use special states, such as temporary immune suppression, to identify those infections that are exacerbated concurrently with exacerbations of the chronic disease in question. Such correlations will then need to be tested for causation by treatment of the exacerbated infection to determine whether the suppression of the infection is associated with amelioration of the disease.

Why will this process change things? For those of us who live in prosperous countries, infectious causes are implicated but not accepted in most common lethal diseases: cancers, heart attacks, stroke, Alzheimer's disease. Infectious causes are also implicated in the vast majority of nonlethal incapacitating illness of uncertain cause, such as arthritis, fibromyalgia, and Crohn's disease. If infectious causes of these diseases are identified, medical history tells us that they will tend to be resolved.

A reasonable estimate of the net effect would be a rise in healthy life expectancy by two or three decades, pushing life span up against the ultimate boundary of longevity molded by natural selection—probably an age range between ninety-five and a hundred and five years. People could thus be expected to live healthy lives into their nineties and then go downhill quickly. This demographic transition toward healthy survival will improve productivity, lower medical costs, and enhance quality of life. In short, it will be one of medicine's greatest contributions.

BIOLOGICAL MARKERS FOR MENTAL ILLNESS

ERIC KANDEL

ERIC KANDEL is a neuroscientist and University Professor and Kavli Professor at Columbia University. He received the Nobel Prize in physiology or medicine in 2000 and is the author of *In Search of Memory: The Emergence of a New Science of Mind*.

Biology in general, and the biology of mind in particular, have become powerful scientific disciplines. But a major lack in the current science of mind is a satisfactory understanding of the biological basis of almost any mental illness. Achieving a biological understanding of schizophrenia, manic-depressive illness, unipolar depression, anxiety states, or obsessional disorders would be a paradigm shift for the biology of mind. It would not only inform us about some of the most devastating diseases of humankind, but, since these are diseases of thought and feeling, it would also tell us more about who we are and how we function.

To illustrate the embarrassing lack of science in this area, let me put this problem into a historical perspective with two personal introductory comments.

First, in the 1960s, when I was a psychiatric resident at the Massachusetts Mental Health Center of the Harvard Medical School, most psychiatrists thought that the social determinants of behavior were completely independent of the biological determinants and that each acted on different aspects of mind. Psychiatric illnesses were therefore classified into two major groups based on presumed differences in origin: organic mental illnesses and functional mental illnesses. That classification, which dated to the nineteenth century, emerged from postmortem examinations of the brains of mental patients. The methods available for examining the brain at that time were too limited to detect subtle anatomical changes; thus only mental disorders that entailed significant loss of nerve cells and brain tissue, such as Alzheimer's disease, Huntington's disease, and chronic alcoholism, were classified as organic diseases and were thought to be based on biology. Schizophrenia, the various forms of depression, and the anxiety states produced no readily detectable loss of nerve cells or other obvious changes in brain anatomy and therefore were classified as functional—that is, not based on biology. Often a special social stigma was attached to the so-called functional mental illnesses, because they were said to be "all in a patient's mind." This notion was accompanied by the suggestion that the illness may have been put into the patient's mind by his or her parents.

With the passage of forty years, we have made progress and seen the advent of a paradigm shift for the science of mind. We no longer think that only certain diseases affect mental states through biological changes in the brain; indeed, the underlying precept of the new science of mind is that all mental processes are biological. They all depend on molecules and cellular processes that occur literally "in our heads"; therefore, any disorder or alteration of those processes must have a biological basis.

In 2001, the late Max Cowan and I were asked to write a review for the *Journal of the American Medical Association* about molecular-biological contributions to neurology. In writing the review, we were struck by the radical way in which molecular genetics had transformed neurology. This led me to wonder why molecular biology has not had a similar transformative effect on psychiatry. The fundamental reason is that neurological diseases and psychiatric diseases differ in several important ways.

Neurology has long been based on the knowledge of where in the brain specific diseases are located. The diseases that form the central concern of neurology—strokes, tumors, and degenerative diseases of the brain—produce clearly discernible structural damage in the brain. Studies of those disorders taught us that in neurology, location is key. We have known for almost a century that Huntington's disease is a disorder of the caudate nucleus of the brain, Parkinson's disease is a disorder of the substantia nigra, and amyotrophic lateral sclerosis (ALS) is a disorder of motor neurons. We know that each of these diseases produces its distinctive disturbances of movement because each involves a different component of the motor system. In addition, such common neurological illnesses as Huntington's, the fragile-X form of mental retardation, some forms of ALS, and the early-onset form of Alzheimer's were found to be inherited in a relatively straightforward way, implying that each of these diseases is caused by a single defective gene.

Pinpointing the genes and defining the mutations that produce these diseases therefore has been relatively easy. Once a mutation is identified, it becomes possible to express the mutant gene in worms, flies, and mice and thereby to discover its mechanism of pathogenesis (how the gene gives rise to disease).

Thus, over the last twenty years, neurology has been revolutionized by the advent of molecular genetics. Because once the anatomical location, identity, and mechanism of action of specific genes are known, diagnoses of neurological disorders are no longer based solely on behavioral symptoms. Indeed, neurologists have even established new diagnostic categories for the neurological diseases, such as the ion channelopathies (for instance, familial periodic paralysis, characterized by aberrant function of ion-channel proteins) and the trinucleotide repeat disorders (for instance, Huntington's disease and fragile-X syndrome, where there is an abnormal and unstable replication of short repeating elements in DNA, which alters the function of the resulting protein). These new diagnostic categories are based not on symptomatology but on the dysfunction of specific genes, proteins, neuronal organelles, or neuronal systems. Moreover, molecular genetics has given us insight into the mechanisms of pathogenesis of neurological disease—insight that did not exist twenty years ago. Thus, in addition to examining patients in the office, physicians can examine brain scans to see how specific regions have been affected by a disorder and order tests for the dysfunction of specific genes, proteins, and nerve cell components.

By contrast to its brilliant illumination of neurology, molecular genetics has so far had only minor effects on psychiatry. We may well ask why.

Tracing the causes of mental illness is a much more difficult task than locating structural damage in the brain. The same factors that have proved useful in studying neurological illnesses have been limiting in the study of mental illness. A century of postmortem studies of the brains of mentally ill persons has failed to reveal the clear, localized lesions seen in neurological illness. Moreover, psychiatric illnesses are disturbances of higher mental function. The anxiety states and the various forms of depression are disorders of emotion; schizophrenia is a disorder of thought. Emotion and thinking are complex mental processes mediated by complex neural circuitry. Until quite recently, little was known about the neural circuits involved in normal thought and emotion.

Furthermore, although most mental illnesses have an important genetic component, they do not have straightforward inheritance patterns, because they are not caused by mutations of a single gene. There appears to be no single gene for schizophrenia or for anxiety disorders, depression, or most other mental illnesses. Instead, the genetic components of these diseases are thought to arise from interaction of several genes with the environment. Each of these genes is thought to exert a relatively small effect; together, however, they create a genetic predisposition for a disorder. Most psychiatric disorders are caused by a combination of these genetic predispositions and some additional, environmental factors. For example, identical twins have identical genes. If one twin has Huntington's disease, so will the other. But if one twin has schizophrenia, the other has only a 50 percent chance of developing it. To trigger schizophrenia, some nongenetic factors in early life—such as intrauterine infection, malnutrition, stress, or the sperm of an elderly father—are required. Because of this complexity in the pattern of inheritance, we have not yet identified most of the genes involved in the major mental illnesses and we have no satisfactory animal models for most mental disorders.

What then is needed to achieve a better biological understanding of mental illness? Two initial requirements are essential and in principle obtainable within the next two decades:

(1) We need *biological markers for mental illness* so that we can understand their anatomical basis, diagnose them objectively, and follow their response to treatment. A beginning is evident in the case of depression, which is associated with hyperactivity in Brodmann Area 25 of the prefrontal cortex; in anxiety states, where there is hyperactivity in the amygdala; and in obsessive-compulsive neurosis, where there is an abnormality in the striatum.

(2) We need *identification of the genes* for various mental illnesses, so that we can understand the molecular basis of these diseases.

These two advances would enhance our ability to understand these disorders better and recognize them earlier. And they would open up new approaches to the treatment of mental illness, an area that has been at a pharmacological standstill for depression, bipolar disorders, and schizophrenia for more than twenty years.

RECOGNIZING THAT THE BODY IS NOT A MACHINE

———◇———

RANDOLPH NESSE

RANDOLPH NESSE is a psychiatrist at the University of Michigan and the coauthor, with George C. Williams, of *Why We Get Sick: The New Science of Darwinian Medicine*.

Many people think that genetic engineering will change everything, even our very bodies and minds. It will, eventually. Right now, however, attempts to apply new genetic knowledge are having profound effects—not on our bodies but on how we understand our bodies. These attempts reveal that our central metaphor for the body is fundamentally flawed. The body is not a machine; it is something very different, a soma shaped by selection, with systems unlike anything an engineer would design. Replacing the machine metaphor with a more biological view of the body will change biology in fundamental ways.

The transition will be difficult, because the metaphor of body-as-machine has served us well. It sped escape from vitalism and encouraged analyses of the body's components, connections, and functions as if they were the creations of some extraordinarily clever cosmic engineer. It has yielded explanations with boxes and arrows, as if the parts are components of an efficient device. Thanks to the metaphor of the body-as-machine, vitalism has been replaced by a deep understanding of the body's mechanisms.

Now, however, genetic advances are revealing the metaphor's limitations. For instance, a decade ago it was reasonable to think we would find the genes that cause bipolar disease. New data have dashed these hopes. Bipolar disease is not caused by consistent genetic variations with large effects; instead, it may arise from many different mutations or from the interacting tiny effects of dozens of genes.

We like to think of genes as information quanta, whose proteins serve specific functions. However, many genes regulate the expression of other genes, which regulate developmental pathways, which are also regulated by environmental factors, which are detected by yet other bodily systems unlike those in any machine. Even the word "regulate" implies coherent planning, when the reality is systems that work, one way or another, by mechanisms sometimes so entangled we cannot fully describe them. We can identify the main players—insulin and glucagon in glucose regulation, the amygdala in responding to threats and losses. But the details? Dozens of genes, hormones, and neural pathways influence one another in interactions that defy description, even while they do what needs to be done. We have assumed, following the metaphor of the machine, that the body is extremely complex. We have yet to acknowledge that some evolved systems may be *indescribably* complex.

Indescribable complexity implies nothing supernatural; bodies and their origins are purely physical. It also has nothing to do with so-called irreducible complexity, that last bastion of creationists desperate to avoid the reality of unintelligent design. Indescribable complexity does, however, confront us with the inadequacy of models built to suit our human preferences for discrete categories, specific functions, and one-directional causal arrows. Worse than merely inadequate, attempts to describe the body as a machine foster inaccurate oversimplifications. Some bodily systems cannot be described in terms simple enough to be satisfying; others may not be described adequately even by the most complex models we can imagine.

This does not mean we should throw up our hands. Moving to a more fully evolutionary view of organisms will improve our understanding. The foundation is recognizing that the body is not a machine. I like to imagine the body as a Rube Goldberg device, modified by generations of blind tinkerers, with indistinctly separate parts connected not by a few strings and pulleys but by myriads of mechanisms interacting in ways no engineer would tolerate, let alone imagine. But even this metaphor is flawed. A body is a body is a body. As we come to recognize that bodies are bodies, not machines, everything will change.

THE ORGANISM ITSELF AS THE EMERGENT MEANING

◄◇►

BRIAN GOODWIN

The late BRIAN GOODWIN was a biologist at Schumacher College, UK, and the author of *How the Leopard Changed Its Spots: The Evolution of Complexity.*

I anticipate that biology will go through a transforming revelation/revolution like the revolution that happened in physics with the development of quantum mechanics nearly a hundred years ago. In biology, this will involve the realization that to make sense of the complexity of gene activity in development, the prevailing model of local mechanical causality will have to be abandoned. In its place, we will have a model of interactive relationships within gene transcription networks, like the pattern of interactions between words in a language, where ambiguity is essential to the creation of emergent meaning sensitive to cultural history and context. The organism itself is the emergent meaning of the developmental process, an embodied form sensitive to both historical constraint within the genome and to environmental context—as we see in the adaptive creativity of evolution. Contemporary studies have revealed that genes are not independent units of information that can be transferred between organisms to alter phenotypes, but elements of complex networks acting together in a morphogenetic process that produces coherent form and function as embodied meaning.

A major consequence of this revelation in biology is the realization that the separation we have made between human creativity (as expressed in culture) and natural creativity (as expressed in evolution) is mistaken. The two are more deeply related than previously recognized. Humans are embedded in and dependent on nature; this has become all too evident recently, as our economic system has collapsed along with the collapse of many crucial ecosystems, because of our failure to integrate human economic activity as a sustainable part of Gaian regulatory networks. We now face dramatic changes in the climate that will require equally dramatic changes in our technologies connected with energy generation, farming, travel, and lifestyle in general.

The recognition that culture is embedded in nature is not so evident, but it will, I believe, emerge as part of the biological revelation/revolution. Biologists will realize that all life, from bacteria to humans, involves a creative process grounded in natural languages as the foundation of its capacity for self-generation and continuous adaptive transformation. The complexity of the molecular networks regulating gene activity in organisms reveals a structure and a dynamic that have the self-similar characteristics and long-range order of languages. The coherent form of an organism emerges during its development as the embodied meaning of the historical genetic text, created through the process of resolving ambiguity and multiple possibilities of form into appropriate functional order that reflects sensitivity to context. Such use of language in all its manifestations in the arts and the sciences is the essence of cultural creativity.

I see the deep conceptual changes currently happening in biology as a prelude and accompaniment to the changes happening in culture, facilitating these and ushering in a new age of sustainable living on the planet.

FASTER EVOLUTION MEANS MORE ETHNIC DIFFERENCES

―――◦―――

JONATHAN HAIDT

JONATHAN HAIDT is a professor of psychology at the University of Virginia and the author of *The Happiness Hypothesis: Finding Modern Truth in Ancient Wisdom*.

The most offensive idea in all of science for the last forty years has been the possibility that behavioral differences between racial and ethnic groups have some genetic basis. Knowing nothing but the long-term offensiveness of this idea, a betting person would have to predict that as we decode the genomes of people around the world, we're going to find deeper differences than most scientists now expect. Expectations, after all, are not based purely on current evidence; they are biased, even if only slightly, by the gut feelings of the researchers, and those gut feelings include disgust toward racism.

A wall has long protected respectable evolutionary inquiry from accusations of aiding and abetting racism. That wall is the belief that genetic change happens at such a glacial pace that there simply was not time, in the fifty thousand years since humans spread out from Africa, for selection pressures to have altered the genome in anything but the most trivial way (for example, changes in skin color and nose shape as adaptive responses to cold climates). Evolutionary psychology has therefore focused on the Pleistocene era—the period from about 1.8 million years ago to the dawn of agriculture—during which our common humanity was forged for the hunter-gatherer lifestyle.

But the writing is on the wall. Russian scientists showed in the 1990s that strong selection pressure—picking out and breeding only the tamest fox pups in each generation—created what was, in behavior as well as body, essentially a new species in just thirty generations. That would correspond to about seven hundred and fifty years for humans. Humans may never have experienced such a strong selection pressure for such a long period, but they surely experienced many weaker selection pressures that lasted far longer and for which some heritable personality traits were more adaptive than others. It stands to reason that local populations (not continent-wide "races") adapted to local circumstances by a process known as coevolution, in which genes and cultural elements change over time and mutually influence each other. The best documented example of this process is the coevolution of genetic mutations enabling the digestion of lactose in adulthood with the cultural innovation of keeping cattle and drinking their milk. This process has happened several times in the last ten thousand years, not to whole "races" but to tribes or larger groups that domesticated cattle.

Recent surveys of the genome across human populations show that hundreds of genes have been changing during the last five to ten millennia, in response to local selection pressures (see papers by Benjamin Voight, Scott Williamson, and Bruce Lahn). No new mental modules can be created from scratch in a few millennia, but slight tweaks to existing mechanisms can happen quickly, and small genetic changes can have big behavioral effects, as with those Russian foxes. We must therefore begin looking beyond the Pleistocene and turn our attention to the Holocene era as well—the last ten thousand years. This was the period after the spread of agriculture, during which the pace of genetic change sped up in response to the enormous increase in the variety of ways that humans earned their living, formed larger coalitions, fought wars, and competed for resources and mates.

The protective wall is about to come crashing down, and all sorts of uncomfortable claims are going to pour in. Skin color has no moral significance, but traits that led to Darwinian success in one of the many new niches and occupations of Holocene life—traits such as collectivism, clannishness, aggressiveness, docility, or the ability to delay gratification—are often seen as virtues or vices. Virtues are acquired slowly, by practice within a cultural context, but the discovery that there might be ethnically linked genetic variations in the ease with which people can acquire specific virtues is—and this is my prediction—going to be a game-changing scientific event. (By "ethnic," I mean any group of people who believe they share common descent, or actually do share common descent, and that that descent involved at least five hundred years of a sustained selection pressure—such as sheep herding, rice farming, exposure to malaria, or a caste-based social order—that favored some heritable behavioral predispositions and not others.)

I believe that the bell-curve wars of the 1990s over race differences in intelligence will seem genteel and short-lived compared with the coming arguments over ethnic differences in moralized traits. I predict that this war will break out between 2012 and 2017.

There are reasons to hope that we'll ultimately reach a consensus that does not aid and abet racism. I expect that dozens, or hundreds, of ethnic differences will be found, so that any group—like any individual—can be said to have many strengths and a few weaknesses, all of which are context dependent. Furthermore, these cross-group differences are likely to be small when compared with the enormous variation *within* ethnic groups and the enormous and obvious effects of cultural learning. But whatever consensus we ultimately reach, the ways in which we now think about genes, groups, evolution, and ethnicity will be radically changed by the unstoppable progress of the Human Genome Project.

AFRICA

◁◇▷

JAMES J. O'DONNELL

JAMES J. O'DONNELL, a classicist and the provost of Georgetown University, is the author of *The Ruin of the Roman Empire: A New History.*

"Africa" is the short answer to the *Edge* question. But this needs explanation.

Historians can't predict black-swan game changers any better than economists can. An outbreak of plague, a nuclear holocaust, an asteroid on collision course, or just an unassassinated pinchbeck dictator at the helm of a giant military machine—any of those can have transformative effect and will always come as a surprise.

But at a macro level, it's easier to see futures—just hard to time them. The expansion of what my colleague the great environmental historian John McNeill calls "the human web" to build a planetwide network of interdependent societies is simply inevitable, but it's taken a long time. Rome, Persia, and ancient China each built a network of empires stretching from Atlantic to Pacific but never made fruitful contact with one another, and their empire-based model of "globalization" fell apart in late antique times. A religion-based model kicked in then, with Christianity and Islam taking their swings. Those were surprising developments, but they went only so far.

It took until early modern times and the development of new

technologies for a real "worldwide web" of societies to develop. Even then, development was Eurocentric for a very long time. Now, in our time, we've seen one significant game changer. In the last two decades, the Eurocentric model of economic and social development has been swamped by the sudden rise of the great emerging market nations—China, India, and Brazil—and many smaller ones. The hope of my youth—that foreign aid would help the poor nations bootstrap themselves—has come true, sometimes to our thinly veiled disappointment, because we suddenly find ourselves competed with for steel and oil and other resources, suddenly find our products in competition with other economies' output. And we wonder whether we really wanted that game to change after all. The slump we're in now is the inevitable second phase of that expansion of the world community, and the rise that will follow is the inevitable third—we all hope it comes quickly.

But a great reservoir of misery and possibility awaits: Africa. Humankind's first continent and homeland has been relegated for too long to disease, poverty, and sometimes astonishingly bad government. There is real progress in many places, but spectacular failures persist. That can't last. The final question facing humankind's historical development is whether the whole human family, including Africa's near-billion, can all achieve, together, sustainable levels of health and comfort.

When will we know? That's a scary question. One future timeline has us peaking now and subsiding, as we wrestle with the challenges we have made for ourselves, into some long period of not-quite-success, while Africa and the failed states of other continents linger in waiting for—how long? Decades? Centuries? There are no guarantees about the future. But as we think about the financial crises of the present, we have to remember that what is at risk is not merely the comfort and prosperity of the rich nations but the very lives of the poorest.

EPISTEMOLOGY WILL CHANGE THE WORLD

———◁◦▷———

LERA BORODITSKY

LERA BORODITSKY is an assistant professor of psychology, neuroscience, and symbolic systems at Stanford University.

There is an old joke about a physicist, a biologist, and an epistemologist being asked to name the most impressive invention or scientific advance of modern times. The physicist does not hesitate: "It's quantum theory, which has completely transformed the way we understand matter."

The biologist says, "No, it's the discovery of DNA, which has completely transformed the way we understand life."

The epistemologist looks at them both and says, "I think it's the Thermos."

The Thermos? Why on earth the Thermos? "Well," the epistemologist explains patiently, "If you put something cold in it, it will keep it cold. And if you put something hot in it, it will keep it hot."

Yeah, so what? "Aha!" the epistemologist raises a triumphant finger. "How does it know?"

This suggests that it's foolhardy to claim that epistemology will change the world. And yet that is precisely what I intend to do here. I think that knowledge about how we know will change everything. By understanding the mechanisms of how humans create knowledge, we will be able to break through normal human cognitive limitations and think the previously unthinkable.

The reason the change is happening now is that modern cognitive science has taken over the role of empirical epistemology. The empirical approach to the origins of knowledge is bringing about breathtaking breakthroughs and turning what once were age-old philosophical mysteries into mere scientific puzzles.

Let me give you an example. One of the great mysteries of the mind is how we are able to think about things we can never see or touch. How do we come to represent and reason about abstract domains like time, justice, or ideas? All of our experience with the world is physical, accomplished through sensory perception and motor action. Our eyes collect photons reflected by surfaces, our ears receive air vibrations created by physical objects, our noses and tongues collect molecules, and our skin responds to physical pressure. In turn, we are able to exert physical action on the world through our motor responses, bending our knees and flexing our toes just the right amount to defy gravity. And yet our internal mental lives go far beyond what is apprehended through physical experience: We invent sophisticated notions of number and time, we theorize about atoms and invisible forces, and we worry about love, justice, ideas, goals, principles. So, how is it possible for the simple building blocks of perception and action to give rise to our ability to reason about domains like those?

Previous approaches to this question have vexed scholars. Plato, for example, concluded that we cannot learn such things and must instead be recollecting them from past incarnations of our souls. As silly as this answer may seem, it was the best we could do for several thousand years. Even some of our most elegant and modern theories (for example, Chomskyan linguistics) have been awkwardly forced to conclude that highly improbable modern concepts like "carburetor" and "bureaucrat" must be coded into our genes (a small step forward from invoking past incarnations of our souls).

But in the past ten years, research in cognitive science has started uncovering the neural and psychological substrates of abstract thought, tracing the acquisition and consolidation of information from motor movements to abstract notions like mathematics and time. These studies have found that human cognition, even in its most abstract and sophisticated form, is deeply embodied, deeply dependent on the processes and representations underlying perception and motor action. We invent all kinds of complex abstract ideas, but we have to do it with old hardware: machinery that evolved for moving around, eating, and mating, not for playing chess, composing symphonies, inventing particle colliders—or engaging in epistemology, for that matter. Being able to reuse this old machinery for new purposes has allowed us to build tremendously rich knowledge repertoires. But it also means that the evolutionary adaptations related to basic perception and motor action have inadvertently shaped and constrained even our most sophisticated mental efforts. Understanding how our evolved machinery both helps and constrains us in creating knowledge will allow us to create new knowledge, either by using our old mental machinery in newer ways or by using new and different machinery for knowledge making, augmenting our normal cognition.

Why will knowing more about how we know change everything? Because everything in our world is based on knowledge. Humans, leaps and bounds beyond any other creatures, acquire, create, share, and transmit vast quantities of knowledge. All scientific advances, inventions, and discoveries are acts of knowledge creation. We owe civilization, culture, science, art, and technology all to our ability to acquire and create knowledge. When we study the mechanics of knowledge building, we are approaching an understanding of what it means to be human—the very essence of human nature. Understanding the building blocks and the limitations of the normal human knowledge-building mechanisms will allow us to get beyond them. And what lies beyond is—well, yet unknown.

SOCIAL MEDIA LITERACY

—◦—

HOWARD RHEINGOLD

HOWARD RHEINGOLD is a communications expert and the author of *Smart Mobs: The Next Social Revolution.*

Social media literacy is going to change many games in unforeseeable ways. Since the advent of the telegraph, the infrastructure for global ubiquitous broadband communication media has been laid down, and of course the great power of the Internet is the democracy of access—in a couple of decades, the number of users has grown from a thousand to a billion. But the next important breakthroughs won't be in hardware or software but in know-how, just as the most important aftereffects of the printing press were not in improved printing technologies but in widespread literacy. The Gutenberg press itself was not enough. Mechanical printing had been invented in Korea and China centuries before the European invention. For a number of reasons, a market for print and the knowledge of how to use the alphabetic code for transmitting knowledge across time and space broke out of the scribal elite that had controlled it for millennia. From around twenty thousand handwritten books in Gutenberg's lifetime, the number of books grew to tens of millions within decades of the invention of movable type. And the rapidly expanding literate population in Europe began to create science, democracy, and the foundations of the Industrial Revolution. Today, we're seeing the beginnings of scientific, medical, political, and social revolutions—from the instant epidemiology that broke out online when SARS became known to the world to the use of social media by political campaigns. But we're only in the earliest years of social media literacy. Whether universal access to many-to-many media will lead to explosive scientific and social change depends more on know-how now than on physical infrastructure. Would the early religious petitioners during the English Civil War, and the printers who eagerly fed their need to spread their ideas, have been able to predict that within a few generations monarchs would be replaced by constitutions? Would Roger Bacon and Isaac Newton have dreamed that entire populations, and not just a few privileged geniuses, would aggregate knowledge and turn it into technology? Would those of us who used slow modems to transmit black-and-white text on the early Internet fifteen years ago have been able to foresee YouTube?

THE DECLINE OF TEXT

MARTI HEARST

MARTI HEARST is a computer scientist at UC Berkeley's
School of Information and the author of *Search User Inter-*
faces.

As an academic, I am loath to think about a world without
reading and writing, but with the rapidly increasing ease of recording and distributing video, and its enormous popularity, it is
only a matter of time before text and the written word become
relegated to specialists such as lawyers and hobbyists.

Movies have already replaced books as cultural touchstones
in the United States. And most Americans dislike watching
movies with subtitles. I assume that given a choice, the majority of Americans would prefer a video-dominant world to a text-
dominant one. (I don't feel I can speak for other cultures.) A
recent report by the Pew Research Center included a quote from
a media executive who said that e-mails containing podcasts
were opened 20 percent more often than standard marketing
e-mail. And I was intrigued by the use of YouTube questions in
last year's U.S. presidential debates. Most of the selected citizen-
submitted videos consisted simply of people pointing the camera
at themselves and speaking their question, with a wall in their
house as backdrop. There were no visual flourishes; the videos
did not add much beyond what a questioner in a live audience

would have conveyed. Video is becoming a mundane way to communicate.

Note that I am not predicting the decline of erudition, in the tradition of Allan Bloom. Nor am I suggesting that video will make us stupid, as argued in Neil Postman's 1985 landmark *Amusing Ourselves to Death*. The situation is different today. In Postman's time, the dominant form of video communication was television, which allowed only for one-way, broadcast-style interaction. We should expect different consequences when everyone uses video for multiway communication.

My argument is that the forms of communication that will do the cultural heavy lifting will be audio and video, rather than text. How will this come about? As a first step, there will be a dramatic reduction in keyboarding; input of textual information will move toward audio dictation. (There is the problem of how to avoid disturbing your officemates or exposing your seatmates on public transportation to private, and doubtless unwelcome, information; perhaps some sound-canceling technology will be developed to solve that problem.) This audio approach will succeed where formerly it has failed because of ease-of-use improvements in editing, storage, and retrieval of spoken words and future improvements in speech recognition technology.

There already is robust technology for watching and listening to video at faster-than-recorded speed without undue auditory distortion (Microsoft has an excellent in-house system for this). And, as noted, technology for recording, editing, posting, and storing video has become ubiquitous and easy to use. As for the use of textual media to respond to criticisms and to cite other work, we already see "video responses" as a heavily used feature on YouTube. One can imagine how technology and norms will develop to further enrich this kind of interaction.

The missing piece in such technology is an effective way to search for video content. Automated image analysis is still an unsolved problem, but there may well be a breakthrough on the horizon. Most algorithms of this kind are developed by "training"—that is, by exposing them to large numbers of examples. The algorithms, if fed enough data, can learn to recognize patterns that can be applied to recognize objects in videos the algorithm hasn't yet seen. This kind of technology is behind many of the innovations we see in Web search engines, such as accurate spell-checking and improvements in automated language translation. Not yet available are huge collections of labeled image and video data, where words have been linked to objects within the images, but there are efforts afoot to harness the willing crowds of online volunteers to gather such information.

What about the developing versus the developed nations? There is, of course, an enormous literacy problem in developing nations. Researchers are experimenting with cleverly designed tools such as the Literacy Bridge Talking Book project, which uses a low-cost audio device to help teach reading skills. But perhaps just as developing nations leap-frogged developed ones by skipping landline telephones to go straight to cell phones, the same may happen with skipping written literacy and moving directly to screen literacy.

I am not saying that text will disappear entirely; one countertrend is the replacement of orality with text in certain forms of communication. For short messages, texting is efficient and unobtrusive. And there is the question of how official government proclamations will be recorded. Perhaps there will be a requirement for transliteration into written text, as a provision of the Americans with Disabilities Act, for the hearing-impaired (although we can hope in the future for increasingly advanced technology to reverse such conditions). But I do think the importance of written words will decline dramatically, both in culture and in how the world works. In a few years, will I be submitting my response to the *Edge* question as a podcast?

THE END OF ANALYTIC SCIENCE

MIHALY CSIKSZENTMIHALYI

MIHALY CSIKSZENTMIHALYI is a psychologist, the director of the Quality of Life Research Center at Claremont Graduate University, and the author of *Flow: The Psychology of Optimal Experience.*

The idea that will change the game of knowledge is the realization that it is more important to understand events, objects, and processes in their relationship with one another than in their singular structure.

Western science has achieved wonders with its analytic focus, but it is now time to take synthesis seriously. We shall realize that science cannot be value-free after all. The Doomsday Clock ticking on the cover of the *Bulletin of the Atomic Scientists* ever closer to midnight is just one reminder that knowledge ignorant of consequences is foolishness.

Chemistry that shrugs at pollution is foolishness. Economics that discounts politics and sociology is just as ignorant as are politics and sociology that discount economics.

Unfortunately, it does not seem to be enough to protect the neutral objectivity of each separate science in the hope that the knowledge generated by each will be integrated later, at some higher level, and used wisely. The synthetic principle will have to become a part of the fundamental axioms of each science.

How shall this breakthrough occur? Current systems theories are necessary but not sufficient, as they tend not to take values into account. Perhaps after this realization sets in, we shall have to rewrite science from the ground up.

COORDINATED COMPUTATIONAL POWER WILL CHANGE SCIENCE

———◦———

LISA RANDALL

LISA RANDALL is a physicist at Harvard University and the author of *Warped Passages: Unraveling the Mysteries of the Universe's Hidden Dimensions*.

Predicting the future is notoriously difficult. Toward the end of the nineteenth century, the famous physicist William Thomson, more commonly known as Lord Kelvin, proclaimed the end of physics. Despite the silliness of declaring a field moribund, particularly one that had been subject to so many important developments not long before this pronouncement, you can't really fault the poor devil for not foreseeing quantum mechanics and relativity and the revolutionary impact they would have. Seriously, how could anyone, even someone as smart as Lord Kelvin, have predicted quantum mechanics?

So I'm not going to even try predicting the unknown. I'll stick to a safer and more prosaic prediction that has already begun its realization. Increases in computing power, in part through shared computational resources, are likely to transform the nature of science and further revolutionize the spread of information. Individual computing power might increase according to Moore's Law, but a more discrete jump in computational power should also result from clever uses of computers in concert. Already we have seen SETI embark on a large-scale search for extraterrestrial signals—a search that would not be possible with any individual computer. Protein folding is now being studied through a distributed computational effort.

Currently, CERN, the European Organization for Nuclear Research, is developing "grid computing," to allow the increase in computational power that will be required to analyze the enormous amount of data from the Large Hadron Collider (LHC). Though the grid system would be hard-pressed to achieve the transformative power of the World Wide Web (also developed at CERN), the jump in computational power possible with processors coordinated in the way that data are can have enormous transformational consequences.

Modern science has two different streams that face very different challenges. Physicists and biologists today, for example, ask very different sorts of questions and use somewhat different methods. Traditionally, scientists have searched for the smallest and most basic components from which the behavior of large complex systems can be derived. This mode has been extremely successful in understanding and interpreting the physical world. For example, it has helped us understand the operation of the human body. I'm betting that this reductionist approach will continue to work for some fields of science, such as particle physics.

However, understanding some of the complex systems that modern science now studies is unlikely to be so simple. Although the LHC's search for more fundamental building blocks is likely to be rewarded with deeper understanding of the substructure of matter, it is not obvious that the most basic structure of biological systems will be understood with as straightforward a reductionist approach.

Very likely, individual elements will work in conjunction with their environment or in collaboration with other system elements to produce emergent effects. Already we have learned that the genetic code is not sufficient to predict behavior; environments that determine which genes are triggered also play a big role. Very likely, understanding the brain will require understanding coordinated dynamics as much as any individual element. Many diseases, too, are unlikely to be completely cured until the complicated dynamics among different elements is fully processed.

How can massive computing power affect such science? It will clearly not replace experiments or the need to identify individual fundamental elements. But it will make us better able to understand systems and how elements work in conjunction. Massive simulation "experiments" will help determine how feedback loops work and how any individual element works in concert with the system as a whole. Such experiments will also help determine when current data are insufficient, in that systems are more sensitive to individual elements than anticipated. Computation alone will not solve problems—the full creativity of scientific minds will still be needed—but computational advances will allow researchers to explore hypotheses efficiently.

At a broader level (although one that will affect science, too) coordinated and expanded computational power will also allow a greatly expanded use of the huge amounts of underutilized information currently available. Searching is likely to become a more refined process, wherein one can ask for particular types of data more finely honed to one's needs. Imagine how much faster and easier Googling could be in a world where you "feel lucky" every time (or at least significantly more often).

The advance I am suggesting isn't a revolution, it's simply an adiabatic evolution of advances that are occurring now. But when one asks what science will be like in twenty years, coordinated computation is likely to be one of the factors that will change many things—though not necessarily everything.

CARNICULTURE

◄◦►

AUSTIN DACEY

AUSTIN DACEY is a philosopher with the Center for Inquiry, a think tank concerned with the secular, scientific outlook, on the editorial staff of *Skeptical Inquirer* and *Free Inquiry* magazines, and the author of *The Secular Conscience: Why Belief Belongs in Public Life*.

Nobody eats animals—not the whole things. Most of us eat animal parts, with a few memorable culinary exceptions. And as we become more aware of the costs of meat—to our health, to our environments, and to the lives of the beings we consume—many of us wish to imagine the pieces apart from the wholes. The meat market obliges. It serves up slices disembodied, drained, and reassembled behind plastic, psychically sealed off from the syringe, saw blade, effluent pool, and all the other instruments of so-called husbandry. But of course this is just cynical illusion.

Imagine, though, that the illusion could come true. Imagine giving in to the human weakness for flesh, but without the growth hormones, the avian flu, the untold millions tortured and gone; imagine the voluptuous tenderness of muscle, finally freed from brutality. You are thinking of cultured meat, or in-vitro meat, and already it is becoming technologically feasible.

Research on several promising tissue-engineering techniques led by scientists in the Netherlands and the United States has been accelerating since 2000, when NASA cultured goldfish meat as possible sustenance on space missions. Soon it will be within our means to stop farming animals and start growing meat. Call it carniculture.

With the coming of carniculture (a term found in science-fiction literature, although, etymologically speaking, "carneculture" might be more correct), meat and other animal products can be made safe, nutritious, economical, energy-efficient, and above all, morally defensible. While carniculture may not change everything in the same way agriculture changed everything, certainly it will transform our economy and our relationship to animals.

Grains once roamed free on untamed plains, tomatoes were wild berries in the Andes. And meat once grew on animals.

EXPLOITABILITY

———◇———

DAVID M. BUSS

DAVID M. BUSS is a psychologist at the University of Texas, Austin, and the author of *The Murderer Next Door: Why the Mind Is Designed to Kill*.

Game-changing scientific breakthroughs will come with the discovery of evolved psychological circuits for exploiting other humans—through cheating, free-riding, mugging, robbing, sexually deceiving, sexually assaulting, physically abusing, cuckolding, mate poaching, stalking, and murdering. Scientists will discover that these exploitative resource acquisition adaptations contain specific design features that monitor statistically reliable cues to exploitable victims and opportunities.

Convicted muggers who are shown videotapes of people walking down a New York City street show strong consensus about whom they would choose as a mugging victim. Chosen victims emit nonverbal cues, such as an uncoordinated gait or a stride too short or long for their height, indicative of ease of victimization. These potential victims are high on *muggability*. Similarly, short stride length, shyness, and physical attractiveness provide reliable cues to *sexual assaultability*. Future scientific breakthroughs will identify the psychological circuits of exploiters sensitive to victims who give off cues to cheatability, deceivability, rapeability, abusability, mate poachability, cuckoldability, stalkability, and killability and to groups that emanate cues to free-rideability and vanquishability.

This knowledge will offer the potential for developing novel defenses that reduce cheating, mugging, raping, robbing, stalking, mate poaching, murdering, and warfare. On the other hand, because adaptations for exploitation coevolve in response to defenses against exploitation, selection may favor the evolution of additional adaptations that circumvent these defenses.

Because evolution by selection is a relatively slow process, the acquisition of scientific knowledge about adaptations for exploitation may enable staying one step ahead of exploiters and effectively short-circuit their strategies. Some classes of crime will be curtailed. Cultural evolution, however, being fleeter than organic evolution, may enable the rapid circumvention of antiexploitation defenses. Defenses, in turn, favor novel strategies of exploitation. Dissemination of discoveries about adaptations for exploitation and coevolved defenses may change permanently the nature of social interaction. Or perhaps, like some coevolutionary arms races, these discoveries ultimately may change nothing at all.

POST-RATIONAL ECONOMIC MAN

—◁◦▷—

DAVID BERREBY

DAVID BERREBY is a science writer and the author of *Us and Them: The Science of Identity.*

Global, twenty-first-century society depends on an eighteenth-century worldview. It's an Enlightenment-era model that says the essence of humanity—and our best guide in life—is cool, conscious reason. Though many have noted, here and elsewhere, that this is a poor account of the mind, the rationalist picture still sustains institutions that, in turn, shape our daily lives.

It is because we are rational that governments guarantee our human rights: To "use one's understanding without guidance" (Immanuel Kant's definition of enlightenment), one needs freedom to inquire, think, and speak. Rationality is the reason for elections (because governments not chosen by thoughtful, evidence-weighing citizens would be irrational). Criminal justice systems assume that impartial justice is possible, which means they assume judges and juries can reason their way through a case. Our medical system assumes that drugs work for biochemical reasons applicable to all human bodies—and not that the price on the pill bottle makes a difference in its effectiveness.

And free markets presume that all players are avatars of Rational Economic Man, he who consciously and consistently perceives his own interests, relates those to possible actions, reasons his way through the options, and then acts according to his calculations. When Adam Smith famously wrote that butchers, brewers, and bakers worked efficiently out of "regard for their own interest," he was doing more than asserting that self-interest could be good. He was also asserting that self-interest—a long-lasting, fact-based, explicit sense of "what's good for me"—is possible.

The rationalist model also suffuses modern culture. Rationalist politics requires tolerance for diversity—we can't reason together if we agree on everything. Rationalist economics teaches the same lesson: If everyone agrees on the proper price for all stocks on the market, then there's no reason for those brokers to go to work. This tolerance for diversity makes it impossible to unite society under a single creed or tradition, and that has the effect of elevating the authority of the scientific method. Data, collected and interpreted according to rigorous standards, elucidating material causes and effects, have become our lingua franca. Our modern notions of the unity of humanity are not premised on God or tribe, but on research results. We say, "We all share the same genes" or "We are all working with the same evolved human nature," or we appeal in some other way to scientific findings. This rationalist framework is so deeply embedded in modern life that its enemies speak in its language even when they violate its tenets: Those who loathe the theory of evolution felt obligated to come up with "creation science." Businesses proclaim their devotion to the free market even as they ask governments to interfere with its workings. Then, too, tyrants who take the trouble to rig elections only prove that elections are now a universal standard.

So that's where the world stands today: with banks, governments, medical systems, nation states resting, explicitly or implicitly, on this notion that human beings are rational deciders.

And, of course, this model looks to be quite wrong. That fact is not what changes everything, but it's a step in that direction.

What's killing Rational Economic Man is an accumulation

of scientific evidence suggesting that people have (a) strong built-in biases that make it almost impossible to separate information's logical essentials from the manner and setting in which the information is presented; and (b) a penchant for changing their beliefs and preferences according to their surroundings, social setting, mood, or simply some random fact they happened to have noticed. The notion that "I" can "know" consistently what my "preferences" are—this is in trouble. (I won't elaborate the case against the rationalist model as recently made by, among others, Gary Marcus, Dan Ariely, and Cass Sunstein, because it has been well covered in the recent *Edge* colloquium on behavioral economics.)

What changes everything is not this ongoing intellectual event but the next one: In the next ten or fifteen years, after the burial of Rational Economic Man, neuroscientists and people from the behavioral disciplines will converge on a better model of human decision making. I think it will picture people as inconsistent, unconscious, biased, malleable corks bobbing on a sea of fast-changing influences, and the consequences of that will be huge for our sense of personhood (to say nothing of sales tricks and marketer manipulations).

But I think the biggest shocks might come to, and through, institutions organized on rationalist premises. If we accept the premise that people are highly influenced by other people and by their immediate circumstances, then what becomes of our idea of impartial justice? (Harvard law professor Jon Hanson has been working on that question for some time.) How do we understand and protect democracy, now that Jonah Berger of Wharton has shown that voters are more likely to support education spending just because they happen to cast their ballots in a school? What are we to make of election results, after we've realized that voters have, at best, "incoherent, inconsistent, disorganized positions on issues," as University of Michigan political scientist William Jacoby puts it? How do you assess a town hall debate once you know that people are more tolerant of a new idea if they're sitting in a tidy room than if they're in a messy one?

How do we understand medical care, now that we know that chemically identical pills affect people who think the pills are expensive differently from the way they affect people who think they're cheap? How should we structure markets after learning that even MBAs can be nudged to see $7.00 per item as a fair price just by exposing them to the number 7 a few minutes before? What do we do about standardized testing, when we know that women who are reminded that they're female score worse on a math test than women reminded instead that they are elite college students?

Perhaps we need a new Adam Smith to reconcile our political, economic, and social institutions with our current knowledge of human nature. In any event, I expect to see the arrival of Post–Rational Economic (and Political and Psychological) Humanity. And I do expect that that will change everything.

NOTHING WILL CHANGE EVERYTHING

---◦---

RICHARD FOREMAN

RICHARD FOREMAN is a playwright and the founder and director of the Ontological-Hysteric Theater.

The belief that there is anything that will change things in and of itself stymies, I believe, real change. To believe that anything "will change things" focuses one on the superficial surfaces of things, which indeed change all the time. Such changes—which have occurred and will continue to occur—create an orientation of consciousness that focuses always on the future. But I propose that the only thing that will "change everything" is the refusal to think about the future. And this, of course, is almost impossible for most human beings to do. Therefore, nothing will change everything. (I admit that I myself have fallen prey to this unavoidable human tendency, having written of the future on the *Edge* Web site, proposing that the Internet is creating, and will radicalize in the future, wide-ranging yet depthless "pancake people.")

But if we could stop thinking about the future, the present moment would obviously expand and become the full (and very different) universe.

One can say, "Ah, but this is the animal state." I would answer no, the animal achieves this state automatically, whereas the human being who achieves it does so only by erecting it on a foundational superstructure that postulates a necessary "future" (past based), much as Freud (and others before him) postulated a necessary unconscious out of which the conscious human being emerged.

So a human being able to not think about the future would have become a nonanimal inhabiting the pure present (the dream of so-called avant-garde art, by the way). And animals do not (apparently) make avant-garde art.

Take John Brockman's offered example of a future event that changes everything: Through genetic manipulation, "your dog becomes your cat" (and by implication, I could become you, and so on). I say that this changes only the shell. Such alterations and achievements, along with many others similarly imaginable, add but another room onto the "home" inhabited by human beings, who will still spend most of their time thinking about the future. And nothing, at the deepest level, therefore will ever change a postulated "everything"—not as long as we keep imagining possible "change" that only reinforces the psychic dwelling of our unchanging selves in a "future" that is always imaginary and beyond us.

BEYOND BOOLEAN LOGIC, DIGITAL MANIPULATIONS, AND NUMERICAL EVALUATIONS

---◇---

VERENA HUBER-DYSON

VERENA HUBER-DYSON is a mathematician and professor emerita in the Philosophy Department of the University of Calgary, Alberta, Canada; she is the author of *Goedel's Theorems: A Workbook on Formalization*.

What will change everything is a radical paradigm shift in the scientific method that opens up horizons beyond the reach of Boolean logic, digital manipulations, and numerical evaluations. Due to my advanced age, I am not likely to witness the change, but I am seeing signs and have my hunches. These I will briefly spell out.

To change everything, a radical paradigm shift must interrupt the scientific method's race: *stop* for a moment's reflection. What are you up to?

How do you know your dog would rather be a cat? Just because you prefer cats? Did you ask it? Have you figured out how to ask it?

Having figured out how to do something is not reason enough to actually do it. That's one aspect of the paradigm shift I am expecting—coming from inside the ranks. Evaluation of scientific results and their potential effects on the world as we know it is of particular urgency these days, when news spreads so easily all over the globe. Of course, we do not want to regress to a system of classified information that generates elitism. Well, this problem is creating the not-so-new-anymore philosophical discipline of applied ethics; if only it stays scientifically well informed and focused on concrete issues. The goal of this part of the shift is a tightening of the structure of the whole conglomerate of the sciences and their presentation in the media.

But this brings me to the more radical effect of the shift I am envisaging: a healing of the rift between the endeavors labeled "scientific" and the proliferation of so-called alternative enterprises, many of which are striving to achieve the blessings of scientific grounding by experimentation, theories, and statistical evaluation, whether appropriate or not—a true and fruitful symbiosis leading to a deeper understanding of the meaning of human existence than as mechanical models or tokens created by a Superior Being for the mysterious purpose of suffering through life in the service of His glory.

Where do I expect the decisive push to come from? Possibly from the young discipline of cognitive science, provided that the disciplines of psychology, philosophy, and physiology are ready to cooperate. There are shoots rising up all over, but I won't embark on a list. Once the "real thing" is found or constructed, it will be recognized. It will have shape and make sense.

The myth of the scientific method as the only approach to reality will become obsolete without loss to humanity's interaction with this world. The path to understanding has to be prepared by a direct, still somewhat mysterious approach of hunches and intuitions in addition to direct perceptions and sensations. Moreover, the results of that procedure are useless unless suitably interpreted.

Well, this is as far as I am ready to go with this explanation of a hunch. The alternative to my current vagueness would be rigidity, prone to misinterpretation.

As to my own turf, mathematics, I do not believe there will be any radical change. Mathematics is a rock of a structure, here to stay. Mathematical insights do not change, they become clearer, and dead ends are recognized as such, but what is proved beyond doubt is cumulative. However, methodological changes here are in order, as well as metamathematical and philosophical interpretation of the nature of results. So is the evolution of an ever-more lucid language. I believe we'd do well to focus on mathematical intuitionism as our foundation. Boolean thinking has done its service by now.

To sum up what I am expecting of this paradigm shift: clarification, simplification, and unification of our understanding, and with it the emergence of a more lucidly expressive language conducive to the end of fragmentation of knowledge.

PEOPLE WHO CAN INTUIT IN SIX DIMENSIONS

—◁◦▷—

ROBERT SAPOLSKY

ROBERT SAPOLSKY is a neuroscientist at Stanford University and the author of *Monkeyluv: And Other Essays on Our Lives as Animals*.

We humans are pretty impressive when it comes to being able to extract information, to discern patterns from lots of little itsy-bitsy data points. Take a musician sitting down with a set of instructions on a piece of paper—sheet music—and being able to turn it into patterned sound. And one step further is the very well-trained musician who can sit and read through printed music, even an entire orchestral score, and hear it in his head, and even feel swept up in emotion at various points in the reading. Even more remarkable is the judge in a composition competition, reading through a work she has never heard before, able to turn that novel combination of notes into sounds in her head—sounds that she can judge to be hackneyed and derivative or beautiful and original.

And, obviously, we do this in the scientific realm in a major way. We come to understand how something works by making sense of a bunch of independent variables interacting to generate some endpoint. (Oh, so *that's* how mitochondria have evolved to solve that problem. . . . *That's* what a temperate-zone rain forest does to balance those different environmental forces challenging it. . . . Now I know!) The trouble is that it's getting harder to do that in the life sciences, and this is where something needs to happen that will change everything.

The root of the problem is technology outstripping our ability to really make use of it. This isn't so much about being able to get increasingly reductive biological information. Scientists figured out some time ago how to sequence a gene, identify a mutation, get the crystallographic structure of a protein, measure ion flow through a single channel in a cell. The recent development is that we can get staggeringly large amounts of that type of information. We have sequenced not just genes but our entire human genome. And we can compare it with that of other species, or look at genome-wide differences between human populations or even individuals, or collect information about tens of thousands of different genes. And then we can look at the expression of those genes: which ones are active at which time, in which cell types, in which individuals, in which populations, in which species.

We can do epigenomics whereby, instead of cataloging the genes existing in an individual, we can determine which genes have been modified long term to make it easier or harder to activate them—in each particular cell type. Or we can do proteomics, examining which proteins and in what abundance have been made as the end product of the activation of those genes; or post-translational proteomics, examining how those proteins have been modified to change their functions.

The same ability to generate massive amounts of data has emerged in other realms of the life sciences. For example, it is possible to do near-continuous samplings of blood-glucose levels, producing minute-by-minute determinations, or do ambulatory cardiology, generating heartbeat data around the clock for days from an individual going about her business, or use state-of-the-art electrophysiological techniques to record the electrical activity of scores of individual neurons simultaneously.

So we are poised to be able to do massive genomo-epigenomoproteonomo-glyco-endo-neurono-orooniomic comparisons of the Jonas Brothers with Nelson Mandela with a dinosaur pelvis with Wall•E and thus better understand the nature of life.

The problem, of course, is that we haven't a clue about what to do with that much data. By that, I don't mean "merely" how to store, or quantitatively analyze, or present those data visually. I mean how to really think about them.

You can already see evidence of this problem in the proliferation of microarray papers. (This is the approach where you ask, "In this particular type of tissue, which genes are more active and which less active than usual under this particular circumstance?") In the fanciest versions of this approach, you've got thousands of bits of information at the end. And too often what all this suggests is that the scientists have hit a wall as far as being able to squeeze insight out of their study. For example, the conclusion in such a paper might be: "Eleventy genes are more active under this circumstance, whereas umpteen genes are less active, and that's how things work." Or, maybe the punch line is: "Of those eleventy genes that are more active, an awful lot of them have something to do with, say, metabolism—how about that?" Or (sheepish tone): "So, changes occurred in the activity of eleventy-plus-umpteen different genes, and we don't know what most of them do, but here are three that we do know about and which plausibly have something to do with this circumstance, so we're now going to focus on those three that we already know something about and ignore the rest."

In other words, far too often the technologies have outstripped our abilities to be insightful. We have some crutches— computer graphics allow us to display a three-dimensional scatter plot, rotate it, change it over time. But still, we barely hold on.

Whatever is going to change everything will have to wait for, probably, our grandkids. It will come from their having grown up with games and emergent networks and who-knows-what-else that (obviously) we can't even imagine. They'll be able to navigate that stuff as effortlessly as we troglodytes can change radio stations while we're driving and talking to a passenger. In other words, we're not going to get much out of these vast data sets until we have people who can intuit in six dimensions. And then, watch out.

MASSIVE TECHNOLOGICAL FAILURE

———◦———

DAVID BODANIS

DAVID BODANIS is a writer, business consultant, and the author of *Passionate Minds: Emilie du Châtelet, Voltaire, and the Great Love Affair of the Enlightenment.*

The big one coming up is going to be massive technological failure, so strong that it will undermine faith in science for a generation or more.

It's going to happen because science is expanding at a fast rate, and over the past few centuries the more science we've had, then—albeit with some time lags—the more powerful technology we have had.

That's where the problem will arise. With each technology, the amplitude of its effects—both positive and negative—gets greater. Automobiles, for example, are an early-twentieth-century technology (based on eighteenth- and nineteenth-century science) that caused a certain amount of increased mobility as well as a certain number of traffic deaths. The amount on each side was large, but not so large that the negative effects couldn't be accepted. Even when the negative effects came to be understood to include land-use problems or pollution, those have still generally been considered manageable. There's little desire to terminate all scientific inquiry because of them.

Nuclear power is a mid-twentieth-century technology (based on early-twentieth-century science). Its overall power is greater still, and so is the amplitude of its destructive possibilities. Through good chance, its negative use has so far been restricted to the destruction of two cities. Yet even that led to a great wave of generalized, antiscientific feeling, not least from among the many people who had always felt that it was impious to interfere with the plans of God.

The Internet is in many ways an even more powerful technology (based on early-twentieth-century quantum mechanics and mid-twentieth-century information theory). So far, its problems have been manageable, be they the surveillance of personal activity or viruslike intrusions that interrupt important services. But the Internet will get stronger and more widespread, as will the collaborative and other tools allowing its misuse: The negative effects will be greater still.

Thus the dynamic we face. Science brings magic from the heavens. In the next few decades, clearly, it will get stronger. Yet just as inevitably, some one of its negative amplitudes—be it in harming health or security or something as yet unrecognized— will pass an acceptable threshold. When that happens, society is unlikely to respond with calm guidelines. Instead, there will be blind fury against everything science has done.

HAPPINESS

———◇———

BETSY DEVINE

BETSY DEVINE is a journalist and blogger, and the co-author, with Frank Wilczek, of *Fantastic Realities: 49 Mind Journeys and a Trip to Stockholm*.

In the next five years, policymakers around the world will embrace economic theories (for example, those of Richard Layard) aimed at creating happiness. The Tower of Economic Babble is rubble. Long live the new, improved happiness economics!

Cash-strapped governments will love Layard's theory that high taxes on high earners make everyone happier: They reduce envy in the less fortunate while saving those now super-taxed from their regrettable motivation to overwork. It also makes political sense to turn people's attention from upside-down mortgages and looted pension funds to their more abstract happiness that politicians and/or policymakers claim you can increase.

Just a few ripple effects from the coming high-powered promotion of happiness:

- Research funding will flow to psychologists who seek advances in happiness creation.
- Bookstores will rename self-help sections "happiness sections," then vastly expand them to accommodate hedonic workbooks and gratitude journals in rival formats.

- In public schools, "happiness" will be the new "self-esteem," a sacred concept to which mere educational goals must humbly bow.
- People will pursue happiness for themselves and their children with holy zeal; people whose children or spouses display public unhappiness will feel a heavy burden of guilt and shame.

Will such changes increase general citizen happiness? This question is no longer angels-on-head-of-pin nonsense; researchers now claim good measures for relative happiness. The distraction value alone should benefit most of us. But in the short run, I, at least, would be happy to see that my prediction had come true.

OUR BRAVE NEW MAP OF THE WORLD

———◆◇◆———

CHRISTINE FINN

CHRISTINE FINN is a British journalist and archaeologist and the author of *Artifacts: An Archaeologist's Year in Silicon Valley.*

While our minds have been engaging with intangibles of the virtual world, where will our real bodies be taking us on planet Earth, compassed with these new perspectives?

Today we are fluent at engaging with the "other" side of the world. We chart a paradox of scale, from the extraterrestrial to the international. Those places we call home are both intimately bounded and digitally exposed. Our observation requires a new grammar of both voyager and voyeur.

The generation growing up with both digital maps and terrestrial globes will have the technological means to shake up our orthodoxies at the very moment that we need to be aware of every last sprawling suburb and shifting sandbank.

Our brave new map of the world is evolving as one created by us as individuals, as much as one that is geographically verifiable by a team of scientists. It will continue to be a mixture of landscapes mapped over centuries and of unbounded digital terrains—one charted in well-thumbed atlases in libraries and by global positioning systems prodded by fingers on the go. Armchair travelers gathering souvenirs through technology, knowing no bounds; a geography of personal space, extended by virtual reality; a sense of place plotted by chats over garden fences, as much as instant messaging exchanged between our digital selves across time zones drawn up in the steam age.

What will not change is our love for adventure. We will not lose our propensity to explore beyond our own horizons and to reexplore those at home. Our innate curiosity will be as relevant in the digital age as it was to the early Pacific colonizers, to seventeenth-century merchants heading east from Europe, and to America's West-drawn pioneers.

Our primeval wanderlust will continue in meanderings off the path, and these are as necessary for our physical selves as the daydreams that bring forth innovation. We develop ways to secure our coordinates while also straining at the leash. And our maps move with us.

Dislocation is good. Take a traditional map of the world. Cut it in half. Bring the old oceanic edges together, and look what happens to the Pacific.

Seeing the world differently changes everything.

THE UNMASKING OF TRUE HUMAN NATURE

———◆———

AUBREY DE GREY

AUBREY DE GREY is a gerontologist, chairman and chief science officer of the Methuselah Foundation, and the author of *Ending Aging: The Rejuvenation Breakthroughs That Could Reverse Human Aging in Our Lifetime*.

Since I think I have a fair chance of living long enough to see the defeat of aging, it follows that I expect to live long enough to see many momentous scientific and technological developments. Does one such event stand out? Yes and no.

You don't have to be a futurophile, these days, to have heard of the Singularity. What was once viewed as an oversimplistic extrapolation has now become mainstream: It is almost heterodox in technologically sophisticated circles *not* to take the view that technological progress will accelerate within the next few decades to a rate that, if not actually infinite, will so far exceed our imagination that it is fruitless to attempt to predict what life will be like thereafter.

Which technologies will dominate this march? Surveying the torrent of literature on this topic, we can with reasonable confidence identify three major areas: software, hardware, and wetware. Artificial intelligence researchers will, numerous experts attest, probably build systems that are "recursively self-improving"—that understand their own workings well enough to design improvements to themselves, thereby bootstrapping to a state of ever more unimaginable intellectual performance.

On the hardware side, it is now widely accepted as technically feasible to build structures in which every atom is exactly where we wish it to be. The positioning of each atom will be painstaking, so one might view this as of purely academic interest—if it were not for the prospect of machines that can build copies of themselves. Such "assemblers" have yet to be completely designed, let alone built, but cellular-automata research indicates that the smallest possible assembler is probably quite simple. The advent of such devices would rather thoroughly remove the barrier to practicability that arises from the time it takes to place each atom; exponentially accelerating parallelism is not to be sneezed at.

And finally, when it comes to biology, the development of regenerative medicine to a level of comprehensiveness that can give a few extra decades of healthy life to those who are already in middle age will herald a similarly accelerating sequence of refinements—not necessarily accelerating in terms of the rate at which such therapies are improved, but in the rate at which they diminish our risk of succumbing to aging at any age, as I've described using the concept of "longevity escape velocity."

I don't single out one of these areas as dominant. They're all likely to happen, but all have some way to go before their tipping point, so the time frame for their emergence is highly speculative. Moreover, each of them will hasten the others: Superintelligent computers will advance all technological development, molecular machines will surpass enzymes in their medical versatility, and the defeat of our oldest and most implacable foe—aging—will raise our sights to the point where we will pursue other transformative technologies seriously as a society, rather than leaving them to a few rare visionaries. Thus, any of the three—if they don't just wipe us all out (but I think that is improbable)—could be "the one."

Or . . . none of them. And this is where I return to the Singularity. I'll get to human nature soon, fear not.

When I discuss longevity escape velocity, I am fond of highlighting the history of aviation. It took centuries for the designs of Leonardo (who was arguably not even the first) to evolve far enough to become actually functional, and many confident and smart engineers were proven wrong in the meantime. But once the decisive breakthrough was made, progress was rapid and smooth. I claim that this exemplifies a general difference between fundamental breakthroughs (unpredictable) and incremental refinements (remarkably predictable).

But to make my aviation analogy stick, I need to explain the dramatic *lack* of progress in the past forty years (since Concorde). Where are our flying cars? My answer is clear: We haven't developed them because we couldn't be bothered—an obstacle that is not likely to occur when it comes to postponing aging. Progress accelerates only when given impetus by human motivation. Whether it's national pride, personal greed, or humanitarian concern, something—someone—has to be the engine room.

Which brings me, at last, to human nature. The transformative technologies I have mentioned will, in my view, probably all arrive within the next few decades—a time frame I personally expect to see. And we will use them, directly or indirectly, to address all the other slings and arrows that humanity is heir to. Biotechnology to combat aging will also combat infections; molecular manufacturing to build unprecedentedly powerful machines will also be able to perform geoengineering and prevent hurricanes and earthquakes and global warming; and superintelligent computers will orchestrate these and other technologies to protect us even from cosmic threats such as asteroids and, in relatively short order, nearby supernovae. (Seriously.) Moreover, we will use these technologies to address any irritations of which we are not yet aware but that will surface as today's burdens are lifted from our shoulders. Where will it all end?

You may ask why it should end at all—but it will. Arguably there will come a time when all avenues of technology will, roughly simultaneously, reach the point seen today with aviation: where we are simply not motivated to explore further sophistication in our technology but prefer to focus on enriching our lives using the technology that already exists. Progress will still occur, but fitfully and at a decelerating rather than accelerating rate. Humanity will at that point be in a state of complete satisfaction with its condition, complete identity with its deepest goals. Human nature will at last be revealed.

AND IF THE BIG CHANGE DOESN'T ARRIVE?

—◇—

CARLO ROVELLI

CARLO ROVELLI is a professor of physics at the Centre de Physique Théorique de Luminy, Université de la Méditerranée, Marseille, and the author of *Quantum Gravity*.

I grew up expecting that someday I'd travel to Mars. I expected cancer and the flu—and all illnesses—to be cured; robots to take care of labor; the biochemistry of life to be fully unraveled; damaged organs to be re-created in every hospital, the nations of Earth living prosperously in peace, thanks to new technology; and physics to understand the center of a black hole. I expected great changes that did not come. They may. It is possible for unexpected advances to change everything—it has happened in the past—but it is also possible that they may not.

Maybe I am biased by my own research field, theoretical physics. I grew up in awe of the physics of the second half of the nineteenth century and the first third of the twentieth. What a marvel! The discovery of the electromagnetic field, thermodynamics, special relativity, general relativity, curved spacetimes, quantum mechanics, probability waves, black holes, . . . What a feast! The world transforming before our eyes every ten years; reality becoming more subtle, more beautiful. But what has happened on that scale in the last thirty years? We are not sure. Perhaps not much. Big dreams, like string theory and multiverses—but are they credible? We do not know. Perhaps the same passion that charmed me toward the future has driven large chunks of today's research into useless, dead-end dreams. Maybe not. Maybe we are really coming to understand what happened before the Big Bang (a "Big Bounce"?) and what takes place deep down at the Planck scale. ("Loops"? Space and time losing their meaning?) Let's be open to the possibility that we are getting there. Let's work hard to get there. But let's also be ready to recognize that perhaps our dreams are just that: dreams. Too often I hear that somebody is "on the brink" of the great leap ahead. I now tend to doze off when I hear "on the brink." In physics, for fifteen years I've heard that we are "on the brink of observing supersymmetry." Wake me when we get there.

Maybe what really changes everything is not something all that glamorous. What has really changed everything so far? Here are two examples. Until no more than a couple of centuries ago, 95 percent of humanity worked the countryside as peasants; that is, humanity needed the labor of ninety-five out of a hundred of its members just to feed itself. This left a happy few to do everything else. Today only a few percent of us work the fields. A few are enough to feed themselves and everybody else. So most of us—including me and most probably you, my reader—are free to do something else, participating in constructing the world we inhabit, making it a better one, perhaps. What made this huge change in our lives possible? Chiefly the tractor. This humble rural machine has changed our life perhaps more than the wheel or electricity.

Another example? Hygiene. Our life expectancy has nearly doubled from little more than washing our hands and taking showers. Change often comes from where it is least expected—recall the famous prediction often attributed to IBM chairman Thomas Watson in 1943: "I think there is a world market for maybe five computers."

It is good to dream about big changes, actively seek them, be open to them. Otherwise we are stuck where we are. But let us not be blinded by our hopes. Dreams sometimes succeed, sometime fail; the century just ended has shown us momentous examples of both. Are we able to discern hype from substance? Dolly the cloned sheep may have been scientifically important, but I tend to see her just as a funny-born twin sister: She hasn't changed much in *my* life yet. Will she ever? The answer to the *Edge* question may well be "Nothing."

"EVERYTHING" HAS ALREADY CHANGED!

---◆---

KAI KRAUSE

KAI KRAUSE is a software pioneer and the author of *I Think, . . . There . . . 4am!*

Change.

Why this idea—that any state is not a *good* state unless there is growth, expansion, redesign . . . *change?*

It is deeply embedded in the human psyche to see any situation as mere momentary balance, just waiting for the inevitable change to happen. The one constant is always . . . *change.* However, for the larger scale of human lifetimes, change for change's sake is not a worthy mantra. Sometimes stasis, the actual opposite of change, may be the harder achievement, the trickier challenge, and yet the nobler cause.

The lengthy quest as we meander through the years is, in my view, really all about *quality of life* more than just riches, honors, or power. And there I question the role of society-at-large for me in my own private day-to-day cycles: I am not about to wait for any "new world," no matter how "brave," to guide my path toward a fulfilled life.

Here I differ from some of the learned brains in this distinguished forum. Straightforward answers are found in these pages: Climate catastrophe, extraterrestrial life, and asteroid collisions are interspersed with solutions "from the lab": meddling with genes, conjuring up superintelligence and nanotechnology, waiting for the Singularity of hard AI, fearing insurgent robots, clamoring for infinite human life span.

Big themes, well said, but somehow . . . I end up shrugging my shoulders.

Is it not incredibly obvious that *the real future* has always turned out unimaginably different from *any* of the predictions ever offered? That the predictions going back further than a decade or so have always missed the core nature of the changes by a mile?

The more white-lab-coaty the experts, studies, and symposia, the further from the truth their models have turned out to be: much of them overly simplistic, too linear, and often too anthropocentric, a few of them absurd, and almost all of them lacking actual relevance to the real future. They will *all* be slightly off, diagonally different from any of our premonitions (of course, "slightly off" means "parsecs").

My thesis: No need to invoke the far future to talk about change. *We are smack in the middle of it* and have been for quite a while.

Sandwiched between World Trade Center and world trade centrifuge, this decade, aptly called the Zero Years, has already changed *everything*. The collapse of once-hallowed personal freedoms like mail, phone, or banking secrecy was just the start. Whereas global priorities like AIDS research or space travel were allocated budgets in the low billions, we now poker casually with $10 *trillion* to prop up one industry after the next. The very foundations are quivering.

It's not about politics. Simply: *No one* foresaw ten years ago the magnitude of change we find ourselves in the middle of right now. No need for an Apophis asteroid to blast the planet.

But the idea that "change has already happened" can be just as positive.

It is a near impossibility to define *quality of life*—a deeply personal set of values, judgments, and emotions. But in recent

years there have been breakthroughs, a new sense of empowerment, a new degree of functionality for our tools. Research is
now immensely powerful, fast, cheap, and enjoyable; it used to
be a slow, painful, and expensive chore. To have the entirety of
Britannica, Wikipedia, millions of pages of writings of all the
ages; to find answers to almost any question in seconds; to view
them on a huge, sharp, window-to-the-world screen—a pure joy
to "work."

Images: Memories can be frozen in time on any occasion in
beautiful detail and collected by the hundreds of thousands. The
few cineastic masterpieces humankind has produced—among
the wretched majority of trash—one can now own and watch,
rerun, and freeze-frame.

Music: All I ever cherished is at my direct disposal—tens of
thousands of pieces. Bach's lifetime oeuvre, 160 CDs' worth, in
my pocket even! Consider that just a few generations ago.

The Berlin Philharmonic plays NOW, just for ME, exactly
THAT . . . and will even *pause when I pee.*

A perfect cup of tea, the right bread, with great jam.

What more does anyone need?

Billions of our predecessors would have *spontaneously com-
busted* with instant-happiness-overdose syndrome, given all these
wonderful means—and I am not even mentioning heated rooms
lit at night, clean showers, safe food, ubiquitous mobility, or den-
tal anesthesia.

"Changing everything" should really equate to bringing *this
basic state* to the billions of our coinhabitants on this dystopic
dirtball. Now! An end to the endless, senseless suffering is the
most meaningful goal. Yes, indeed, *let's change everything.*

Will I see that in my lifetime? Well, I'll see the beginning.
And there is an optimistic streak in me, hopeful about the human
spirit's ability to face seemingly insurmountable challenges.

Photonic solar paint that gathers free energy on any surface;
transparent photovoltaic film to make every window in any house

or vehicle into a steady energy source; general voltaics with 90 percent efficiency; splitting water into hydrogen with ease (and then using that to power a desalination machine, which in turn can feed the water . . .); transmitting power wirelessly (as MIT finally did, following Tesla's cue). Thin-air "residual energy" batteries powering laptops and cell phones nearly infinitely. All that is very close and will make an enormous difference, not just in Manhattan, London, and Tokyo but also in Ulan Bator, Irian Jaya, and the Pantanal.

No need to invoke grand sweeping forces, momentous upheavals, those armies of nanotech, gene-spliced AI robots. . . .

Let's embrace the peace and quiet of keeping things just as they are for a while. Bring them to the rest of the planet, taking the time to truly enjoy them, milking the moment for all it has, really watching, listening, smelling, and tasting it all.

. . . that *stasis*. . .

That has changed everything.

THE SLOW-MOTION REVOLUTION

————◦————

ROBERT R. PROVINE

ROBERT R. PROVINE is a psychologist and neuroscientist at the University of Maryland and the author of *Laughter: A Scientific Investigation.*

The survival of our ancestors on the savannah depended on their ability to detect change. Change is where the action is. Our nervous system is biased for the detection of change. You don't need to know that things are the same, the same, the same. Do you feel the watch on your wrist or the ring on your finger? Probably not, unless you have just put them on. You don't see the blind spot of each retina, because these areas are unchanging and filled in by your brain with information from the visual surround; if the image on your retina is experimentally stabilized, the entire visual field fades in a few seconds and you can see only visual stimuli that move through the field of view. You notice the sound of your home's air control system when it turns on or turns off, but not when it's running.

Our perception of changing stimulus amplitude is usually nonlinear. The sensation of loudness grows much more slowly (exponent of 0.60) than the amplitude of the physical stimulus, a reason why rock bands have huge amplifiers and speakers. Perceived brightness grows even more slowly than loudness (exponent of 0.33). The sensation of electric shock grows at an accelerating rate (exponent of 3.50), quickly shifting from a just-detectable tingle to an agonizing jolt. Our estimate of length grows linearly (exponent of 1.00); a two-inch line appears twice as long as a one-inch line. We are lousy sound, light, and volt meters but halfway decent measuring sticks.

We are poor at making absolute judgments of stimulus amplitude, basing our decisions on relative, ever-changing standards. We judge ourselves to be warm or cool relative to "physiological zero," our adaptation level. The same room can seem either warm or cool, depending on whether you entered it from a chilly basement or an overheated sunroom. The lesson of temperature judgment is applicable to other, more complex measures of change, associated with wealth and success. For a highly paid CEO, this year's $10-million bonus does not feel as good as last year's bonus of the same size. The second term of a presidency doubtless feels less momentous than the first.

Exploring how we perceive changes in anything suggests the difficulty of identifying something that changes *everything*, from the perspective of the individual. The velocity of change is also critical. Did the Renaissance, the Reformation, the Industrial Revolution, or the Computer Revolution have ordinary people amazed at the changes in their lives? Historical and futuristic speculation about events that change everything features time compression and overestimates the rate of cultural and psychological change. As with previous generations, we may be missing the slow-motion revolution taking place around us, unaware that we are part of an event that will change everything. What is it?

WHY HUMAN NATURE WILL REBEL

—◇—

NICHOLAS HUMPHREY

NICHOLAS HUMPHREY is School Professor at the Centre for Philosophy of Natural and Social Science at the London School of Economics and the author of *Seeing Red: A Study in Consciousness.*

We're easily seduced by the idea that once the Big One comes, nothing will ever be the same again. But I guess what will surprise—and no doubt frustrate—those who dream of a scientifically-driven new order is how unchangeable, and unmanageable by technology, human lives are.

Imagine if this *Edge* question had been posed to the citizens of Rome two thousand years ago. Would they have been able to predict the coming of the Internet, DNA fingerprinting, mind-control, space travel? Of course not. Would that mean they would have failed to spot the technological developments that were destined to change everything? I don't think so. For the fact is, nothing has changed everything.

Those Romans, despite their technological privations, led lives remarkably like ours. Bring them into the twenty-first century and they would of course be amazed by what science has achieved. Yet they would soon discover that beneath the modern wrapping it is business as usual. Politics, crime, love, religion, heroism. The stuff of human biography. The more it changes, the more it's the same thing.

The one development that really could change everything would be a radical, genetically programmed alteration of human nature. It hasn't happened in historical times, and I'd bet it won't be happening in the near future either. Cultural and technical innovations can certainly alter the trajectory of individual human lives. But while human beings continue to reproduce by having sex and each new generation goes back to square one, then every baby begins life with a set of inherited dispositions and instincts that evolved in the technological dark ages.

The Latin poet Horace wrote, "You can drive out nature with a pitchfork, but she will always return." Let's dream, if we like, of revolution. But be prepared for more of the same.